C. Shelton

Electrical installations

Second edition

 Longman

Pearson Education Limited
Edinburgh Gate, Harlow,
Essex CM20 2JE, England
and Associated Companies throughout the world.

First published 1993
Second edition 1996
Reprinted 1999, 2000

ISBN 0-582-273226

British Library Cataloguing in Publication Data
A CIP record for this book is available from the British Library.

Set by 4 in 10/12 pt Compugraphic Times
Produced by Pearson Education Asia Pte Ltd
Printed in Singapore (CNC)

Contents

Preface

Most people will agree that we all try to achieve a good working knowledge of our chosen career and to carry out the task before us with maximum efficiency coupled with the minimum of difficulty. This is an aim we should all try to accomplish, whichever type of electrical work is undertaken. Regulation 130-01-01 confirms.

To the trainee, apprentice or young electrician on the threshold of his or her career, the future can seem daunting, as there may be many unknown facets and unforeseen pitfalls that can affect the outcome of an average job. Insufficient time, bad planning, the wrong materials, hazardous infrastructure and inexperience are some of the possible difficulties to be faced. However, with a little wise thought most problems may be overcome.

Throughout this book I have tried to face squarely some of the more commonplace problems which often arise from time to time in electrical installation and maintenance work. Many of the snags and complications which may have to be confronted have been outlined, and practical solutions suggested to ease these problems.

Previous knowledge in understanding the fundamental concepts of electrical theory is helpful but less desirable than a practical awareness of electrical installation work. Details have been presented in a practical and straightforward manner using everyday language. Mathematics has been kept to a minimum and used only when explaining certain electrical concepts occurring in both the theoretical and practical aspects of electrical installation work.

In Part 1 the theoretical aspects of electrical work are considered, keeping in mind the practicalities involved. In Part 2 a close look is taken at the practical side of the skill. Also explained are methods and techniques needed in order to develop and achieve a greater working knowledge of the task in hand. Part 3 has been written with health and safety in mind and outlines the many hazards and dangers electrical operatives may encounter in the course of a working day.

The knowledge gained from this book will provide valuable insight enabling the advancement of basic electrical skills. From this greater understanding will come the satisfaction of being able to tackle a job with confidence within the electrical installation industry.

The second edition of *Electrical Installations* has been revised to meet the demands of the amended version of the *City and Guilds GCLI 236 Electrical Installation Competences and Works Syllabus*. The book now incorporates knowledge evidence requirements for *National Vocational Qualification* standards (levels 1, 2 and 3), for the electrical installation engineering industry and has been written to satisfy the requirements of the *16th Edition of the IEE Wiring Regulations*.

Graphical symbols used throughout the book are generally drawn from *British Standards 3939* but where clarity is necessary, pictograms are provided and used as an alternative to provide clearer understanding. This eliminates need to make reference to *British Standards* directive 3939.

Christopher Shelton September 1995

Acknowledgements

I would like to thank the following individuals for their assistance and time given in the preparation of this book:

Carl Wollaston for photographic work; Yvonne Palmer for editorial typing; John Dix for Figure 11.6; Paul King for help in producing Figure 16.1; Christopher Barr for illustrative work on Figures 16.3 & 16.4; Brian Field of Access International Limited for help and advice in compiling material relating to free-standing access towers (Access International Limited is a member of the Pre-fabricated Aluminium Scaffolding Manufacturers Association (PASMA)).

I would also like to thank the following companies and establishments for their help and for permission to use copyright material:

AEI Cables Limited; Broadlands Estate, Romsey; Dowding and Mills (Southern) Limited; Institute for Electrical and Electronic Engineers, Inc., Piscataway, New Jersey, USA; Institute of Electrical Engineers; Robin Electronics Limited; Southampton and Salisbury College; Systems and Electrical Supplies Limited; Walsall Conduits Limited; Wiltshire Fire Brigade; Wylex Limited.

Extracts from ANSI Y32.9-1972 are reproduced with the permission of the American National Standards Institute.

Extracts from BS 3939 are reproduced with the permission of BSI. Complete copies can be obtained by post from BSI Customer Services, 389 Chiswick High Road, London W4 4AL; Telephone: 0181 996 7000.

Special thanks to my wife, Shirley, who acted as my literary advisor and typist, for her invaluable support, assistance and encouragement.

Abbreviations

Abbreviations used in mathematical expressions

a   = area
C = correction factor
C_1 = value of capacitor
C_t = total capacitance in farads or microfarads
C_x = value of an unknown capacitor
E = voltage (alternating)
E_p = primary voltage (transformer)
E_s = secondary voltage (transformer)
E_1 = auxiliary electrode voltage
f = frequency in hertz
I = current in amps
I_a = armature current in amps or potential earth fault current
I_p = current in amps (transformer primary winding)
I_s = current in amps (transformer secondary winding)
I_t = total current in amps
K   = degrees Kelvin
K = mathematical constant
L = value of the inductor in henries
m   = metres
N = number of (N is also Neutral)
N_p = number of turns of wire (transformer primary winding)
N_s = number of turns of wire (transformer secondary winding)
p = pairs of poles
Q = capacitance in coulombs
r = internal resistance in ohms*
r^2 = radius, squared
R = resistance in ohms
R_a = armature resistance in ohms
R_f = fault resistance in ohms
R_l = load resistance in ohms
R_t = total resistance in ohms
R_1 = value of consumer's phase conductor in ohms (also used for resistors in series)
R_2 = value of consumer's protective conductor in ohms (also used for resistors in series)
S1   = supplementary earth electrode
t = temperature in degrees Celsius also used for time
t^o = time and degrees
U_0 = nominal voltage
V = voltage (direct current)
W = power in watts
X_C = capacitative reactance in ohms
X_L = inductive reactance in ohms
Z = impedance in ohms
Z_e = earth loop impedance offered by the service cable
Z_s = total earth fault loop impedance in ohms
α = temperature coefficient in ohms per degree Celsius (lower-case Greek letter alpha)
Π = mathematical constant: 3.1416 (Greek capital letter pi)
ρ = resistivity in ohms per metre (Greek lower-case letter rhò)
ϕ = diameter (phi)

Abbreviations used throughout the manuscript

AC   alternating current
Ag   silver
BC   before Christ
BS   British standard
ć   common
c_a correction factor for ambient temperature
Cd   cadmium
c_f correction factor given for semi-closed fuses
c_g correction factor for cable grouping
c_i correction factor for thermal insulation
°C   degrees Celsius
cpc   circuit protective conductor

DB	distribution board*	n	neutron
DC	direct current	NV	no volt*
DP	double pole	O	oxygen
ELCB	earth leakage circuit breaker	PME	protective multiple earthing
emf	electromotive force	PNE	phase, neutral and earth*
e^-	electron*	PVC	polyvinyl chloride
EE	electrode under test	p^+	proton*
f	fuse*	R	resistance*
°F	degrees Fahrenheit	RCD	residual current device
FVCB	fault voltage circuit breaker	RFI	radio frequency interference
H	hydrogen	RSJ	reinforced steel joist*
Hg	mercury (the element)	rpm	revolutions per minute
Hz	frequency in hertz	rps	revolutions per second
H_2SO_4	sulphuric acid	SA	supply authority*
I_B	design current	SELV	safety extra low voltage
IEC	International Electrotechnical Commission	STRAW	stop, think, review and work
IEE	Institution of Electrical Engineers	SWA	stranded wire armoured cable
I_f	earth fault current	T	test*
I_n	current rating for a protective device	t	time in seconds
jb	joint box	uPVC	unplasticised polyvinyl chloride
l	length in metres	VD	volt drop
LSF	low smoke and fumes	VIR	vulcanised india rubber
LNE	live, neutral and earth*	Wh	Watt hour meter
m	metres	=	equal to*
MCB	miniature circuit breaker	+	positive (direct current)*
MCCB	moulded case circuit breaker	−	negative (direct current)*
MI	mineral insulated	μF	microfarads (lower-case Greek letter mu)
mA	milliamp*	$\simeq$	approximately equal to
MICC	mineral insulated copper cable	Ω	ohms (capital Greek letter omega)

* Abbreviations used in figures.

Part 1 Theory with practice in mind

1 Basic atomic theory and fundamental electrical principles

In this chapter: Atomic theory, electromotive force, current flow and the coulomb. The relationship between the ampere, ohm, volt and watt.

Man has questioned the very nature of matter since the beginning of recorded history. It was in 402 BC that the early Greeks cautiously offered to an unsuspecting world their first atomic theory. They concluded that all matter comprised stable, minute substances which were completely indivisible. These particles they called *atoms*.

Atoms are the basic building blocks of the universe, and all matter is made from a combination of them. To understand electricity we must first have a working knowledge of the atom, an understanding of how it is constructed and its relationship with other atoms: a relationship which, under defined conditions, can lead to the phenomenon we loosely know as electricity.

The atomic interior

To familiarise ourselves with the internal construction of the smallest particle of matter we must first take a close look at the hydrogen atom, the lightest of all known elements. An isotope of hydrogen is made up from three principal parts as Figure 1.1 shows. The proton, a positively charged particle, together with the neutron, a particle which carries no charge at all, form what is known as the nucleus of the atom. Our atomic model is completed with a negatively charged electron orbiting the nucleus in a similar fashion to the Moon travelling around the Earth. This orbiting *negatively* charged particle is drawn to the *positively* charged proton, but a constant orbit is continually maintained through a partnership of centrifugal force and mutual attraction. The faster

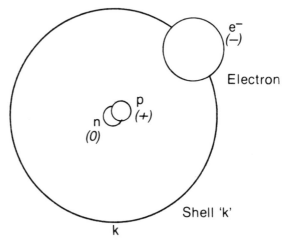

Figure 1.1. An isotope of hydrogen. The neutron (n) has no electrical charge and a mass slightly greater than the positively charged proton (p). The electron (e^-) carries a negative charge.

an electron encircles its atomic nucleus, the further away from the proton it will be, whereas an electron placed near the central mass will be travelling around it at a far lower speed.

Generally, an electron is about three times larger than the proton it orbits. Surprisingly, protons are 1840 times heavier than their orbiting cousins, but whatever the element the basic building blocks of nature remain the same. Only their integral quantities differ.

The atomic nucleus has been calculated to be a 10 000th the size of the complete atom and can be roughly compared to a large marble 10 millimetres in diameter and centrally suspended in a spherical bubble approximately 215 millimetres in diameter. Clearly the vast majority of the volume of the atom is taken up with orbiting electrons and space, as Figure 1.2 shows.

Take for example an atom of silver, illustrated as Figure 1.3. The atom is constructed from 47

3

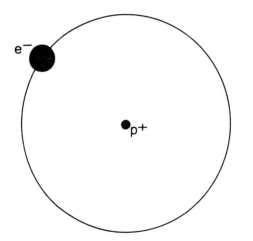

Figure 1.2. An atom of hydrogen (H). The vast majority of the atom is taken up with an orbiting electron and space.

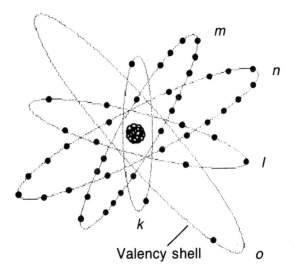

Figure 1.3. An atom of silver (Ag).

protons and 61 neutrons to form the central nucleus. To balance electrically, so that the atom is neither positively nor negatively charged, 47 orbiting electrons are positioned in shells or quanta around the nucleus. The electrons are placed in well-defined orbits which are designated by the lower-case letters *k, l, m, n* and *o*. The shell nearest the central mass is known as the '*k*' shell and contains two electrons. The next shell, or quanta, '*l*' contains 8 orbiting electrons whereas '*m*' and '*n*' shells both accommodate 18 electrons. The last shell '*o*', called the *valency shell*, has just

one orbiting electron. Figure 1.4 shows, in schematic form, the positioning of the electrons in an atom of silver, clarifying how they are arranged in shells or quanta around the nucleus.

The valency shell must be regarded as the most important one in our atomic model for it is in this shell in which electrons are loosely held and allowed to flow from one atom to another whenever an electromotive force is applied. Silver is a good conductor of electricity. It has large atoms and many orbits, as Figure 1.4 illustrates and because the valence electron is much further away from the centre of the atom, there is far less hold on the nucleus. Therefore when an electron is moved by electromotive force from its orbit and strikes a valence electron serving another atom, more energy is produced and so on.

In practical terms, problems will develop when the electron flow is disproportionate to the size of conductor used. Energy, in the form of heat, is released whenever a valence electron knocks another from its orbit. Should the electron flow be excessive, the medium in which they are flowing will proportionally increase in temperature, causing an eventual breakdown to the system.

Should a material have two valence electrons in its outer shell, for example nickel or mercury, the energy level from the striking electrons will be divided. This can be clearly demonstrated by aiming a cue ball at two stationary snooker balls and noting that the resulting collision will cause the energy levels of the two target balls to be halved. Whatever the element, there are never more than eight electrons located in its valence shell. As an example, silver and copper each have one, whereas tin has four, and antimony has five.

The vast majority of current-carrying conductors used in electrical engineering are made from materials whose atoms support just one loosely held valence electron located in their outer shell. Insulators such as glass, plastic or dry wood have many more and are tightly held. Think back again to the snooker cue ball aiming this time at, not one, but eight target balls!

The maximum possible electrons per shell
Atoms have a maximum quantity of electrons that may be contained in one shell or quantum. Other

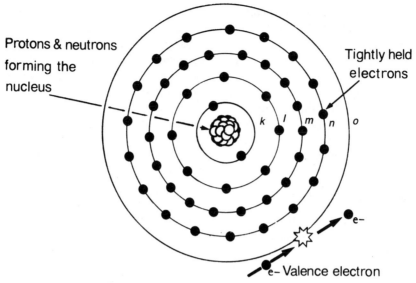

Figure 1.4. An atom of silver: schematic form.

than the valency shell, this number may be calculated by utilising the following expression:

$$(2N^2) \qquad [1.1]$$

where N represents the shell number of the atom.

As an example, consider an atom of mercury (Hg):

Orbit No.	Shell	Maximum number of electrons	Electron configuration
1	k	2	2
2	l	8	8
3	m	18	18
4	n	32	32
5	o	50	18
6	p	8	2

('p' becomes the valency shell)

It is important to remember that there are never any more than eight electrons within the valence quantum, whatever the element might be.

Electromotive force and current flow

The ability to knock loosely held valence electrons from their orbits depends on a suitable prime mover and this could take the form of a simple generator. To demonstrate this principle in very basic terms, two permanent magnets with opposing

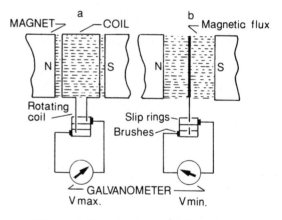

Figure 1.5. The principle of a simple generator.

poles are placed opposite each other as shown in Figure 1.5. A coil of wire is then introduced and positioned at right angles to the magnetic field. At this point no electromotive force (emf) is induced into the coil. When the coil is rotated so that it is seen to be cutting through the invisible magnetic lines of flux, as Figure 1.5(a) shows, a voltage is immediately induced in the coil. This voltage is often referred to as the emf and its magnitude will depend on the velocity of the rotated coil and the density of the magnetic flux.

To summarise this important concept: a coil of wire striking a magnetic field at right angles will induce an emf into it and force loosely held valence electrons to shift from their regular orbits and buffet other outer shell electrons serving neighbouring atoms, *ad infinitum*. This is current flow in its most basic form. Just think of the domino effect! Voltage, on the other hand, does not flow. It is simply a pressure which enables electron flow to be brought about.

The electron, the amp and the coulomb

In order to understand higher concepts of electrical science, we must first familiarise ourselves with the fundamental principles of electron flow, which as we shall see are directly related to current flow. The unit of current is the ampere and it is this we shall be dealing with next.

In the very early days of electrotechnology it was widely thought that electricity flowed from a higher potential to a lower, similar to an elevated water storage tank dispensing water to a tap beneath. History has now taught us that those early concepts were wrong. Electrons, which form the raw material from which electricity is conceived, are negatively charged and flow from the negative to the positive terminal. This is the exact opposite of what was envisaged in those early pioneering days. Rather than completely disrupt the principles that had been previously formulated, we still maintain the conventional hypothesis and say that electricity flows from positive to negative. After all, we do say that the Sun rises in the morning when in fact what we really mean is that the Earth is turning.

Figure 1.6 shows a simple circuit comprising

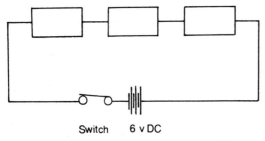

Switch   6 v DC

Figure 1.6. A simple series circuit connected to a 6 V supply.

three 2 ohm resistors wired in series formation and served by a 6 volt direct current supply taken from a battery. The term 'ohm, Ω' is a unit of electrical resistance and may be defined as the resistance between two points of a conductor when a potential of 1 volt, applied between these two points, produces a current of 1 amp within the conductor.

The total resistance of the circuit may be calculated by adding the individual values together. Hence,

$$2 + 2 + 2 = 6\,\Omega$$

Ohm's Law states that the current flowing in a circuit is equal to the voltage applied and inversely proportional to the resistance. Thus,

$$\text{amps} = \frac{\text{voltage}}{\text{resistance}}$$

or   [1.2]

$$I = \frac{V}{R}$$

Substituting figures,

$$I = \frac{6}{6} = 1 \text{ amp}$$

From this calculation it can be seen that the total current flowing in the circuit as illustrated in Figure 1.6 is 1 amp. By definition, an amp is 6.24×10^{18} negative electrons passing a fixed point in a circuit in one second. Looking at it another way, it is the flow of 6 240 000 000 000 000 000 electrons flowing past a given point every second. A daunting task for anyone who wanted to calculate the exact number of electrons flowing in one hour!

We have seen that an amp is represented by many billions of electrons and it is this quantity of electricity passing a fixed point every second that is called a *coulomb (C)*. The value in coulombs (*Q*) for any given circuit may be calculated by use of the following expression:

$$Q = I \times t$$ [1.3]

where Q is the value in coulombs,
    I is the current in amps and
    t is the time period in seconds.

It has been clearly shown that 1 A not only represents 6.24×10^{18} electrons per second, which is quite a mouthful, but is also equal to one coulomb per second. Electron flow is the fundamental principle governing any electrical circuit. Whether large or small, simple or complex, the concept and rules controlling this hypothesis are the same. We shall be dealing with these rules again in Chapters 2 and 3, when we will be discussing Ohm's Law and seeing how the residual current device works in relation to a faulty circuit.

Resistance and power

Resistance

We have learnt that voltage is a term used to describe an electrical potential difference: an electrical pressure that is measured in volts and whose symbol is V. We have also seen how electron flow throughout a conductor is directly related to the current consumed in the circuit and is measured in amps; identified by the capital letter I. Electrical resistance is quantified in *ohms* and symbolised by the letter R and the Greek capital letter omega (Ω).

The ohm is the unit of resistance and is the opposition that is offered to a flow of electrons in a circuit or within a component of a circuit as Figure 1.7 shows. It may be also described as the mathematical ratio of the electrical potential difference between the terminals of a conductor and the current flowing through the conductor.

As a practical example, imagine running up a

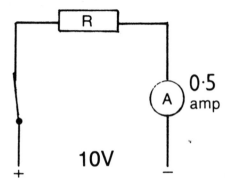

Figure 1.7. The value of an unknown resistor may be calculated by dividing the applied voltage, V, by the current, I.

hill in a small muddy cable ducting littered with scattered fragments of rubbish. It is physical opposition such as this that an electron similarly encounters when negotiating an electrical resistance.

By dividing the total current into the applied voltage, the resistance of the circuit may be found. As an example, consider Figure 1.7:

A carbon resistor of unknown value has been added to complete a simple circuit. By instrumentation the applied voltage was found to be 10 volts and the current 0.5 amps.

By use of the folowing expression the total resistance may be calculated:

$$R = \frac{V}{I} \qquad [1.4]$$

where R is the total resistance in ohms,
V is the applied voltage and
I is the current in amps.

Substituting for known figures,

$$R = \frac{10}{0.5} = 20 \text{ ohms}$$

The unknown resistor is clearly 20 ohms.

Power

Ohm's Law will be studied more fully in Chapter 2, but first the *watt* will be defined, whose symbol is W.

Named after *James Watt* (1736–1819), this fundamental unit of power is defined as *the energy expended by a constant current of one amp flowing through a conductor, the ends of which are maintained at a potential difference of one volt.*

By multiplying the voltage by the current consumed, the power in watts may be calculated. Should there be a need to express in terms of kilowatts (kW), divide the total calculated power in watts by 1000.

Instrumentation has shown that the current flowing in the example served by Figure 1.7 is 0.5 amp and the applied voltage, 10 volts. By use of the following expression, the power in watts may be calculated:

$$W = I \times V \qquad [1.5]$$

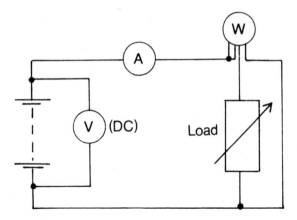

Figure 1.8. Measurement by instrumentation.

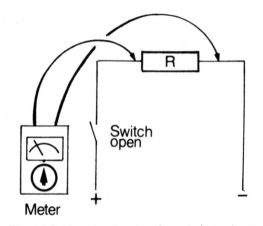

Figure 1.9. Measuring the value of an unknown resistance.

Substituting figures,

$$W = 0.5 \times 10$$
$$W = 5 \text{ watts } (0.005 \text{ kW})$$

Catering for the practical aspects involved, Figure 1.8 describes how current, voltage and power may be measured by instrumentation. Resistance may be measured by use of an ohmmeter, connected as illustrated in Figure 1.9. Remember to isolate the power before carrying out any such test, otherwise possible damage to the meter could result.

Summary

In this chapter we have discussed the concept of the atom and its family of subatomic particles which together form the fundamental building blocks of nature. Under defined conditions, the free electrons loosely held in the outermost shell of the atom will flow and collide with other neighbouring atoms, causing a knock-on effect to produce an electric current flow. We have defined the amp, coulomb and watt in basic terms and examined the relationships existing between these rudimentary units of energy.

It is these that we shall be dealing with in Chapter 2, where the theoretical and practical concepts of Ohm's Law will be discussed more fully.

2 Problems relating to Ohm's Law

In this chapter: Pioneering work. Aspects of power and current. Associated practical problems. Ohm's Law related to primary and secondary cells, steel conduit systems and resistances wired both in series and parallel formation.

One of the first experiments in electricity known to physical scientists was carried out by a Greek philosopher called *Thales* in 550 BC when he produced static electricity by using an amber rod rubbed with fur. Thales reasoned that all matter was comprised of tiny particles called atoms. It was 2000 years later that a Colchester man called *William Gilbert* coined the name 'electricity' when he published his work in 1600 on *The Magnet and Magnetic Bodies*. This subsequently paved the way for all major experiments in the field of electricity.

Georg Simon Ohm, a Bavarian, carried out much of the pioneering work and in 1827 formulated a principle which has been applied ever since. His law is one of the fundamental corner-stones of electrical science and of prime importance to every worker within the electrical industry.

Ohm's Law states that: *The current flowing in a circuit is directly proportional to the voltage applied and inversely proportional to the resistance at a constant temperature.*

For mathematical expedience this is often written as:

$$I = \frac{E}{R} \qquad [2.1]$$

where E is the electromotive force applied (usually an alternating current, AC) or

$$I = \frac{V}{R}$$

where I is the current in amps,
V is the applied voltage (usually a direct

current, DC) and
R is the resistance in ohms.

There are many problems to which Ohm's Law may be applied, so many that it would be possible to fill a large book several times over. The objective of this chapter is to combine theory with practice and to examine some of the more realistic difficulties and problems which may arise in electrical installation work.

Practical problems involving Ohm's Law

The most basic form of Ohm's Law was first introduced to the world by Georg Ohm in 1827 but, as Figure 2.1 shows, there are many different

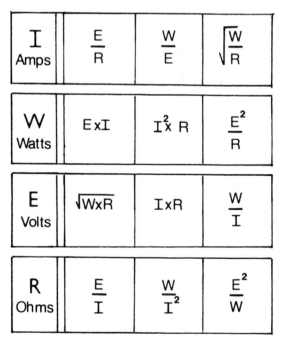

Figure 2.1. Alternative methods of evaluating current, power, voltage and resistance.

variations of this fundamental expression. It would be as well to learn these alternative formulae as Ohm's Law is far from satisfactory should a calculation have to be made to determine the current flowing in a circuit when the only available data are the power generated in watts and the resistance in ohms.

As a practical example, consider the following problem:

A 230 volt heating element of an unknown power output has to be replaced during a planned maintenance programme. Only the voltage and the heating element's resistance can be measured, as other forms of instrumentation are unavailable. Calculate the power generated in watts.

The applied voltage may be measured with the aid of a suitable voltmeter. By use of an ohmmeter, the resistance of the heating element may also be found.

Let us assume the total resistance measured is 26.45 ohms. By referring to Figure 2.1:

$$\text{Power in watts} = \frac{V^2}{R} \qquad [2.2]$$

where V is the applied voltage and
R is the total resistance in ohms.

Substituting symbols for figures,

$$\text{Power in watts} = \frac{230^2}{26.45}$$

$$W = \frac{52\,900}{26.45}$$

$$W = 2000 \text{ watts}$$

The calculation clearly indicates that the power in watts dissipated in the circuit is exactly 2000 (2 kilowatts).

Current and switching problems

A problem that is often discovered after a circuit has been commissioned occurs when an inductive lighting load is controlled by a standard 5 A switch. Within a short space of time the switch will physically deteriorate internally should the connected load be too high. This invariably leads to the switch malfunctioning by burning out.

Take for an example the current drawn from a circuit serving seven twin 85 W fluorescent lighting fittings which are controlled by a 5 A switch from a 230 V supply. Providing the power factor is less than 0.85, the current drawn from the supply would be:

$$I = \frac{\text{total power in watts} \times 1.8}{\text{volts}} \qquad [2.3]$$

Substituting for figures known:

$$I = \frac{(85 \times 2) \times 7 \times 1.8}{230}$$

$$I = \frac{2142}{230}$$

Therefore $I = 9.3$ amps.

Clearly, a 5 A lighting switch would be inadequate for this circuit. It would be far more sensible to install a 20 A switch at the design stage of the installation to maintain a healthier circuit.

If, instead of fluorescent lighting, the circuit had been a *resistive load*, that is to say, incandescent lamps or heating elements, then:

$$I = \frac{85 \times 2 \times 7}{230}$$

$$I = 5.1 \text{ amps}$$

In this case a 5 A switch wired within the circuit would have been satisfactory.

For a practical current evaluation for a circuit serving an *inductive load*, multiply the total wattage by 1.8 before division is carried out.

Assessing current demand

Ohm's Law demonstrates that a current flowing in a circuit is directly proportional to the voltage applied and inversely proportional to the resistance. There are several variations by which this basic law and the relationship between associated terms may be expressed. We shall now deal with one of these variations.

As a practical problem, consider a large open-plan office in the middle of winter, solely relying on one 30 A moulded case circuit breaker (MCCB) to supply a variety of electrical appliances, including heaters. Problems arose when the MCCB intermittently tripped as though experiencing a fault condition. An initial inspection revealed that both the installation and apparatus were in a satisfactory condition.

This then provides us with two major reasons why such a condition could have occurred:

1. Malfunction within the moulded case circuit breaker.
2. An overcurrent.

In an ideal world all electricians would possess a clamp meter enabling them to assess the total current drawn from the final ring circuit. However, if a meter is unavailable, resort must be made to basic theoretical training in order to evaluate the current mathematically.

Figure 2.1 has demonstrated that by multiplying the current flowing in a circuit by the applied voltage, the power in watts may be calculated. Thus:

$$\text{Power in watts} = I \times V \qquad [2.4]$$

Where I is the current in amps,
V is the applied voltage and
W is the power in watts.

Transposing the expression of Equation [2.4] in terms of I by dividing each side of the expression by V:

$$I = \frac{W}{V} \qquad [2.5]$$

By adding the total potential power which might be consumed throughout the circuit and then dividing by the applied voltage, the maximum current may be calculated. This figure will represent the highest possible current demand which might be drawn from the circuit, but in no way reflects the true amount which could be much smaller. Let us suppose that, after calculation, the sum obtained was 15.3318 kW (15 331.8 W).

By applying expression [2.5]:

$$I = \frac{W}{V}$$

the potential current can now be found. Substituting for figures,

$$I = \frac{15\,331.8}{230}$$

$$I = 66.66 \text{ amps}$$

Clearly, 66 A is far too much for the circuit breaker to hold. Even with an assumed 50 per cent diversity, the current flowing in the circuit is well in excess of the 30 A rated overcurrent protection device serving the final ring circuit. The practical answer would be to install an additional circuit in order that the applied load my be distributed more evenly. Reference is made to Regulation 314-01-01.

Problems relating to installation faults in series

Under specific conditions, theoretical complications with practical overtones can occur when carrying out an installation using solid drawn steel conduit. This may be demonstrated, when for example, a joining coupler or brass bush has been defectively installed, creating an unacceptably high impedance in the system.

Figure 2.2 shows a simple, 230 V single phase installation wired in PVC insulated single cable and drawn through steel conduit to serve a 30 A industrial socket outlet, protected by a 30 A enclosed fuse. The installation, controlled by a metal-clad consumer unit is supplied by a 100 A 'TN-S' (see Chapter 8) mains system. Earthing is by means of the service cable's protective sheath. Reference is made to Regulation 413-02-06.

Let the carelessly installed bush have a resistance of 10 Ω measured between the metal

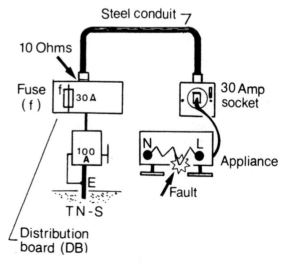

Figure 2.2. Carelessly installed steel conduit can cause many problems should a fault occur.

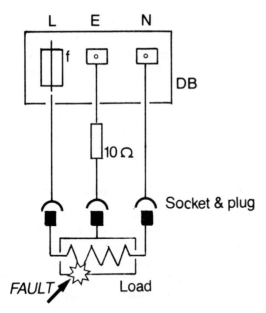

Figure 2.3. The installation fault shown in schematic form.

consumer unit and the conduit. Should a 'live' to earth fault of negligible impedance occur within the appliance, on the side of higher potential, the appliance, socket outlet and conduit would become 'live'. Figure 2.3 clearly shows both primary and secondary fault conditions in schematic form.

By applying Ohm's Law, expression [2.1], it will be seen that the current flowing in the current protective conductor, which in this case is the steel conduit installation, is:

$$\text{amps} = \frac{\text{volts}}{\text{resistance}}$$

Substituting figures,

$$\text{amps} = \frac{230}{10} = 23 \text{ amps}$$

Equating in terms of power, Expression [1.5] shows that the power generated in watts is the product of the current multiplied by the applied voltage, thus:

$$23 \times 230 = 5290 \text{ watts } (5.29 \text{ kW})$$

A situation such as this can cause many problems. Not only would a dangerously high potential exist on all exposed conductive parts, but the power generated in the faulty circuit would be

sufficient to destroy the PVC insulation on the cables within the steel conduit.

This clearly is a potential fire risk and safety hazard of which we must all be aware. The rule must always be to check and tighten thoroughly all metal conduit joints before the installation is commissioned. Remember that a rewirable fuse will maintain nearly twice its rated current before the fuse element ruptures. A great deal of damage could be done.

If in doubt, or unable to provide a suitable earth from the distribution point by way of the steel conduit, add an appropriately sized current protective conductor (cpc) to the installation. Reference is made to Regulation 547-03.

Resistances in series formation

We have seen how a 'live' to earth fault can occur within the framework of a steel conduit system, and how power is generated under such conditions. We will now consider a common problem which can affect both phase and neutral conductors and the connected load.

Faults where two resistances are connected in series formation, where one is the product of a fault condition, can be calculated mathematically. To achieve this, we must first look at the theoretical concepts involved.

Resistances wired in series formation may be calculated by use of the following expression:

$$R_t = R_1 + R_2 + R_3 + R_4 \ldots \qquad [2.6]$$

Where R_t is the total resistance in ohms and
R_1, R_2, R_3, R_4 are the individual resistance values.

To provide a clearer understanding of this concept, consider the following example:

Figure 2.4 illustrates a simple circuit employing three 2 Ω and one 10 Ω resistor wired in series with each other and supplied by an emf of 32 V.

Substituting figures, the total resistance in the circuit may be calculated by use of Expression [2.6]:

$$R_t = 2 + 2 + 2 + 10$$
$$R_t = 16 \text{ ohms}$$

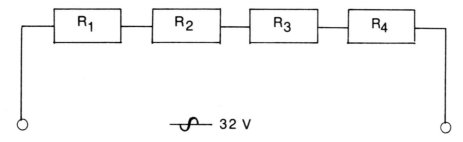

Figure 2.4. Resistances wired in series formation.

By applying Ohm's Law, Expression [2.1], the current flowing in the circuit may also be found. Hence:

$$I = \frac{V}{R}$$

$$I = \frac{32}{16} = 2 \text{ amps}$$

Since all the components are wired in series formation, it stands to reason that the current flow is the same throughout the circuit. Imagine the resistors as an electric fire element.

Volt drop

The individual loads will produce a volt drop (VD) which is in direct proportion to the resistance offered to the voltage applied. To determine the value of the volt drop across a series circuit, the following expression is used:

$$\text{volt drop} = I \times R \qquad [2.7]$$

where I is the total current in amps flowing in the circuit and R is the individual resistance under examination. From Figure 2.4, using Expression 2.7:

volt drop across resistor $R_1 = 2 \times 2 = 4$ volts

volt drop across resistor $R_2 = 2 \times 2 = 4$ volts

volt drop across resistor $R_3 = 2 \times 2 = 4$ volts

volt drop across resistor $R_4 = 2 \times 10 = 20$ volts

By adding the respective volt drops together, the total applied voltage may be calculated. Hence:

Applied voltage $= \text{VD}_1 + \text{VD}_2 + \text{VD}_3 + \text{VD}_4$
$\qquad\qquad\qquad\qquad\qquad\qquad\qquad [2.8]$

Substituting figures,

$$V = 4 + 4 + 4 + 20$$
$$V = 32 \text{ volts}$$

The power in watts may be calculated in a similar manner as determining the volt drop across each resistor by applying the following expression:

$$W = I^2 \times R \qquad [2.9]$$

where W is the power in watts,
$\qquad I$ is the current in amps and
$\qquad R$ is the individual resistances in ohms.

A practical example of this expression may be shown when an electrical appliance is served through a faulty line connection:

Let us suppose a 230 volt electric fire offering a total resistance of 20 ohms, is supplied through a badly connected terminal serving a fused connection unit. The fault, illustrated in Figure 2.5, has a measured resistance of 8.75 Ω between the incoming and outgoing supply cables.

Referring to Expression [2.6]:

$$R_t = R_1 + R_2$$

FAULT 8.75Ω

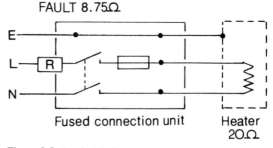

Figure 2.5. An electric heater served by a poorly connected fused connection unit.

or, suiting the needs of the problem,

$$R_t = R_f + R_l \qquad [2.10]$$

where R_f is the resistance of the fault condition in ohms and
R_l the total resistance of the load in ohms.

Substituting figures,

$$R_t = 8.75 + 20 = 28.75 \text{ ohms}$$

This effectively reduces the power output of the electric fire by nearly a third.

The total current may now be calculated by applying Ohm's Law, Expression [2.1]:

$$I = \frac{V}{R}$$

$$I = \frac{230}{20 + 8.75} = 8 \text{ amps}$$

The volt drop across the fault condition may also be found by using the expression:

$$I \times R_f = \text{volt drop (VD)} \qquad [2.11]$$

and, substituting figures,

$$VD = 8 \times 8.75 = 70 \text{ volts}$$

Similarly, the volt drop across the load may be defined as:

$$VD = 8 \times 20 = 160 \text{ volts}$$

By adding the two volt drops together, a check may be made, as the answer will correlate with the supply voltage.

Expression [2.9] affirms that:

$$\text{Power in watts} = I^2R$$

Substituting figures and calculating the power dissipated by the fault condition:

$$W = 8^2 \times 8.75$$
$$W = 560 \text{ watts}$$

(enough to do considerable damage) and the power dissipated by the load will be:

$$W = 8^2 \times 20$$
$$W = 1280 \text{ watts}$$

Figure 2.6 shows, in schematic detail, the relationship between the current flowing in the

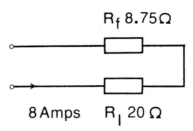

Figure 2.6. Schematic relationship between current flowing and the impedance offered by the load and fault condition.

circuit and the impedance offered by the load and fault condition. In practice, an eventual breakdown would occur, rendering considerable damage to both the fused connection unit and the supporting circuitry. It is important to ensure that all cables are secure to avoid any possible fault condition occurring due to badly terminated wiring. Regulation 526 confirms.

Ohm's Law and primary and secondary cells

It can be extremely irritating when rechargeable batteries no longer hold their charge, or completely lose their effectiveness through no obvious fault caused by the operative. Although not a great deal may be achieved to return the battery to its former glory, it is helpful to understand why cells behave in this way.

Problems involving batteries are very often the direct result of a large internal resistance developing in the cell. This is due to chemical and plate sediment offering a suitable passage to the current from one plate to the other.

If the internal resistance, r, is of a high magnitude, rapid discharge of the cell will result without any external load being connected across its terminals. This will obviously affect the total voltage output of the battery, which could lead to unforeseen external consequences; for example, if the cell was within a bank of batteries serving an emergency lighting arrangement, or supplying a stable source of power to a fire alarm system (Regulation 313-02-01).

Figure 2.7 represents a faulty 24 V lead acid accumulator connected across an external load of 5.1 Ω. When measuring the current drawn from the supply, it was found to be 4 A. Calculation shows this to be incorrect, as Ohm's Law states

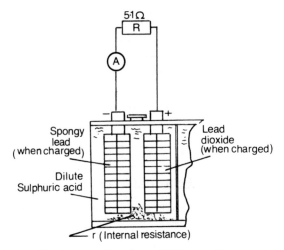

Figure 2.7. A faulty cell serving a 24 V lead acid accumulator.

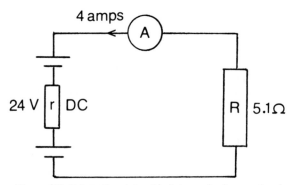

Figure 2.8. Schematic relationship between load, current and the internal resistance of a faulty cell.

that the current flowing in a circuit is the ratio of voltage and resistance, or:

$$\text{current in amps} = \frac{\text{voltage}}{\text{resistance}} \qquad [1.2]$$

Substituting figures,

$$I = \frac{24}{5.1} = 4.705 \text{ amps}$$

The current reading of 4 A, illustrated in Figure 2.8, must be the direct result of the internal resistance of the cell. Both internal and external loads are in series formation and therefore will proportionally affect the current flowing throughout the circuit.

By use of the following expression, the value of the internal resistance, r, may be established:

$$I = \frac{V}{R + r} \qquad [2.12]$$

where I is the current in amps,
V is the terminal voltage of the battery,
R is the external load in ohms and
r is the internal resistance of the cell.

First, cross-multiply and present the expression in terms of V,

$$V = I(R + r) \qquad [2.13]$$

and then in terms of r,

$$r = \frac{V}{I} - R \qquad [2.14]$$

Substituting known values,

$$r = \frac{24}{4} - 5.1$$

$$r = 0.9 \text{ ohms}$$

By calculation it has been established that the internal resistance, r, is 0.9 Ω.

Another aspect of this problem occurs when the source of power is a portable generator. The internal resistance may then be calculated using the following expression:

$$r = \frac{E - V}{I} \qquad [2.15]$$

where E is the electromotive force generated in volts,
V the value of the terminal voltage under load and
I is the value of the current drawn in amps.

Compound internal resistance
Several batteries, each with a similar internal resistance and wired together in series formation, will cause a compound internal resistance to develop within the cells. Figure 2.9 shows in schematic form six healthy batteries connected in series formation, developing a total output of 12 V.

Suppose an external resistance, R, of 6 Ω is placed across the terminals. By applying Ohm's Law, it may be calculated that the current flow (V/R) is 2 A and the power generated ($I \times V$) is

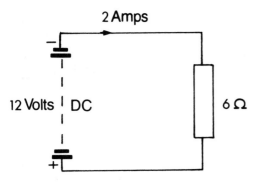

2 Amps

12 Volts DC   6 Ω

Figure 2.9. Six healthy 2 V cells wired in series formation will develop an output of 12 V.

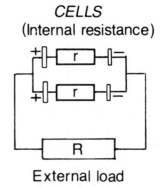

CELLS
(Internal resistance)

External load

Figure 2.11. Two single cells wired in parallel formation, each with similar internal resistances.

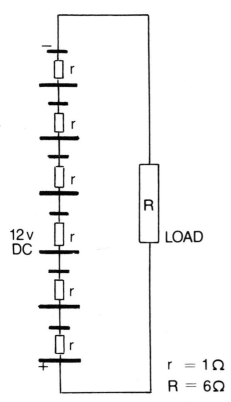

12 v
DC

R LOAD

$r = 1\,\Omega$
$R = 6\,\Omega$

Figure 2.10. Six cells, each with an internal resistance of 1 Ω wired in series formation.

24 W. If the six healthy cells were to be replaced with cells each with an internal resistance of 1 Ω (Figure 2.10), the current flowing would be calculated by use of the following expression:

$$I = \frac{V}{R + r_1 + r_2 + r_3 + r_4 + r_5 + r_6}$$
[2.16]

Substituting figures,

$$I = \frac{12}{6 + 1 + 1 + 1 + 1 + 1 + 1}$$
$$= 1 \text{ amp}$$

and the power in watts $= I \times V$ (expression [1.5]). Substituting figures for known quantities,

$$W = 1 \times 12 = 12 \text{ watts}$$

This is half the original value evaluated from figures drawn from the healthier circuit. A similar group of batteries serving an emergency lighting or fire alarm system would be condemned and replaced immediately.

Figure 2.11 illustrates two one-celled batteries with similar internal resistances, r, connected in parallel formation. When connected to an external resistance, R, the current drawn from the circuit may be found by using the expression:

$$I = \frac{V}{R + r/2}$$
[2.17]

where r is the internal resistance of the cell.

A cell with an internal resistance provides a vital path for electrons to flow between the two opposing plates as Figure 2.12 clearly shows. Current is constantly being drawn from the battery and if left unchecked it will completely run down. A good tip is always to check the health of the accumulator regularly by using a direct current voltmeter and a suitable hydrometer to measure the specific gravity of the acid. This way, any

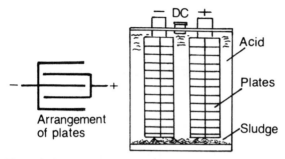

Figure 2.12. A cell with an internal resistance will quickly run down.

problems can be exposed and remedial steps taken before serious complications occur.

Installation problems involving Ohm's Law

Situations often arise from time to time when theory can be called upon to solve practical problems in electrical installation work: problems and difficulties that would otherwise go unanswered during the normal course of events. In this chapter it will be demonstrated how theory may be used in order to solve puzzling electrical problems.

Take as an example a situation that might arise should a customer insist on providing his or her own apparatus and mains supply. Assume the equipment is designed to serve a small greenhouse and is supplied by an old DIY 5 A 230 V single phase supply. The apparatus provided comprises one quartz halogen flood light and a bank of tubular heaters.

Closer inspection shows that although not entirely past its useful life, the apparatus is very old and any reference to load which might once have been present has been erased. It is important at this stage to assemble as much information and data as possible so that the job may be carried out both safely and satisfactorily.

Checking the continuity, measurements indicated that the quartz halogen light had a resistance of 57.6 Ω and the bank of tubular heaters 19.2 Ω.

Since both light and tubular heaters are wired in parallel formation, although separately switched (Figure 2.13), the total resistance in ohms must first be determined in order that the potential current may be calculated.

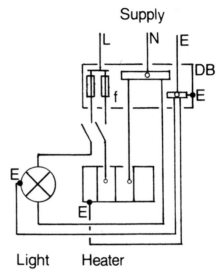

Figure 2.13. Both lighting and heating circuits are wired in parallel formation.

The total resistance of the proposed installation may then be determined by employing the following expression:

$$\frac{1}{R_t} = \frac{1}{R_1} + \frac{1}{R_2} \qquad [2.18]$$

where R_t is the total resistance in ohms and R_1 and R_2 are the resistances of the two loads.

Substituting figures,

$$\frac{1}{R_t} = \frac{1}{57.6} + \frac{1}{19.2}$$

Selecting a common multiple and working through:

$$\frac{1}{R_t} = \frac{19.2 + 57.6}{1105.92}$$

$$\frac{1}{R_t} = \frac{76.8}{1105.92}$$

By cross-multiplying, and bringing the expression in terms of R_t:

$$R_t = \frac{1105.92}{76.8}$$

$$R_t = 14.4 \text{ ohms}$$

By applying Ohm's Law, Expression [2.1], the

potential current flowing in the circuit may be calculated:

$$I = \frac{V}{R}$$

Substuting figures for known values,

$$I = \frac{230}{14.4} = 15.9 \text{ amps}$$

Calculations have shown that the proposed 5 A supply is inadequate to support the envisaged load and in practice a new supply would have to be installed.

A quicker way to calculate the total resistance offered is to employ the following expression:

$$R_t = \frac{R_1 \times R_2}{R_1 + R_2} \qquad [2.19]$$

A word of warning when using this expression: it is only suitable for *two* known resistance values. Should there be more than two, Expression [2.18] should be used to solve the problem.

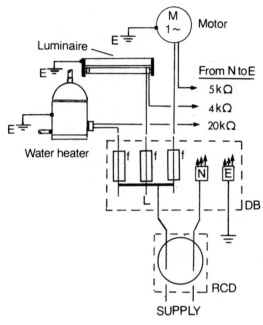

Figure 2.14. Parallel paths between the neutral and current protective conductor will produce a low impedance value.

Multiple faults

Electrical installations employing residual current devices can cause much frustration to the operative should a multiple fault occur, causing the device to activate either constantly or intermittently. This problem will be dealt with more fully in Chapter 3 and also Part 2 of this book.

A residual current device will automatically isolate the supply when, for example, there are three insulation faults, but will remain stable with two faults. In order to understand the reason for this anomaly we must revert once more to theory.

Consider the following practical problem as outlined in Figure 2.14. This is a typical example where parallel and independent resistance paths track between the neutral conductor and earth, producing a very low impedance when measured with an insulation tester.

A recent insulation test carried out on a small-holding indicated three independent neutral to earth faults. Upon inspection it was established that the three paths to earth were 5000, 4000 and 20 000 Ω respectively. The small-holding, supplied by a 230 V single phase supply, is served by a residual current rated at 100 mA.

Independently, neither of the fault conditions would be sufficiently low to activate the 100 mA residual current device. The smallest recorded value of 4000 Ω would only draw 57.5 mA from the supply. Since all three known resistances are in parallel formation (neutral to earth), their total combined resistance will always be less than the smallest value. By using Expression [2.18]. the total resistance of the parallel fault paths may be calculated:

$$\frac{1}{R_t} = \frac{1}{R_1} + \frac{1}{R_2} + \frac{1}{R_3}$$

Substituting figures,

$$\frac{1}{R_t} = \frac{1}{5000} + \frac{1}{4000} + \frac{1}{20\,000}$$

Selecting a common multiple, 20 000,

$$\frac{1}{R_t} = \frac{4 + 5 + 1}{20\,000}$$

$$\frac{1}{R_t} = \frac{10}{20\,000}$$

By cross-multiplying and bringing the equation in terms of R_t,

$$R_t = \frac{20\ 000}{10}$$

$$R_t = 2000 \text{ ohms}$$

The total resistance between the neutral conductor and earth is calculated to be 2000 Ω.

By applying Ohm's Law, Expression [2.1], the total earth leakage current may be calculated:

$$I = \frac{V}{R}$$

Substituting figures:

total current leakage (I) to earth

$$= \frac{230 \text{ volts}}{2000 \text{ ohms}}$$

$$I = 0.115 \text{ amps}$$

By multiplying by 1000 the current in amps may be expressed in terms of milliamps (1000 milliamps = 1 amp):

$$0.115 \text{ amps} \times 1000 = 115 \text{ milliamps}$$

Calculations have shown that the earth leakage current flowing throughout the installation is greater than the residual current device can sustain and in practical terms would trip out.

If confronted with a similar situation where an impedance exists between the neutral and earth conductor, remember that *any* circuit which is drawing current is capable of activating the device. All neutral conductors are common with each other, as illustrated in Figure 2.15. Therefore any healthy circuit drawing current would experience divided neutral paths on its return journey to the supply transformer. The majority of current in the neutral conductor would be returned via the common connection point at the distribution centre but a small amount would leak to earth by way of the fault condition within the installation.

It is this small amount of current leakage which is crucial when trying to maintain the electrical balance of the residual current device. The current entering must be the same value as the current departing from the apparatus. If not, an imbalance

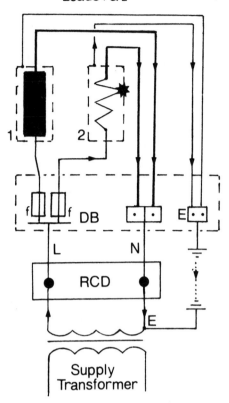

Loads 1 & 2

Figure 2.15. Given that all neutrals are common, a fault occurring with any neutral conductor would experience a divided return path to the supply transformer.

will occur and the device will detect a fault condition and trip.

On a practical note, fault-finding within an installation served by a residual current device is made far easier by the use of a good quality digital continuity meter switched to the ohm or milli-ohm scale. The nearer the test is to the fault, the lower the reading will register on the instrument.

The installation of a residual current device must meet the demands of Regulations 531-02-01 to 531-05-01.

Summary

In this chapter we have considered the work of Georg Simon Ohm and related practical everyday problems to his outstanding work. We have seen that there are many ways in which Ohm's Law may be expressed and how these variations can help solve numerous theoretical complications that can occur in electrical installation work. We have investigated and reviewed internal factors that can influence and effect the output of a secondary cell and how we might relate these to Ohm's Law. Practical difficulties involving circuitry served by residual current devices have been examined and solutions offered.

3 Current- and voltage-operated earth leakage circuit breakers

In this chapter: Theoretical concepts. Practical applications. Nuisance tripping. Disadvantages (current- vs. voltage-operated). Fault-finding techniques.

Nature has provided most of us with many defences in the form of inherited instincts which may be used to protect ourselves against numerous dangers. Regrettably, we do not have the ability to perceive the presence of electricity.

Fortunately, rules and regulations governing the safe use of electricity were laid down as long ago as 1882. Since then, steady progress has been made over the years producing safe, sensible methods and techniques designed to safeguard and protect both electrical worker and consumer alike.

Electrical protection can be offered in many styles ranging from the humble fuse to sophisticated automatic circuit disconnection techniques. With the advancement of new and modern technology, devices have been introduced that will switch off automatically should a fault condition place a dangerously high voltage on any exposed conductive part.

Appliances such as these are called *residual current devices (RCDs)*.

The residual current device

Figure 3.1 illustrates a simple scenario in which a faulty electric fire is served by a 13 A plug whose current protective conductor has been carelessly disconnected. Under normal healthy working conditions, such a break might be unnoticed. However, a fault occurring to earth within the appliance would, under these circumstances, be of serious concern. Had the installation been installed without a residual current device and supplied from a 'TT' distribution system (see Chapter 8),

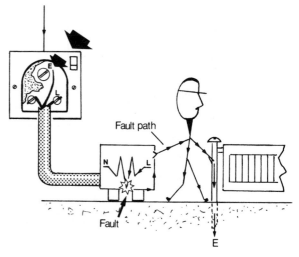

Figure 3.1. Fault path taken in the absence of a current protective conductor.

there would have been little or no safeguard against a fault condition to earth. If an unfortunate victim simultaneously touched the faulty appliance and an extraneous conductive part, he or she could sustain a lethal electric shock.

Figure 3.2 illustrates the basic internal wiring arrangements serving a typical RCD. This means of protection is more likely to be found in rural areas where the local electricity authority's supply is delivered by overhead cables forming what is known as a 'TT' system. This system provides both phase and neutral into the customer's property but an independent earth has to be provided by the consumer. As an effective earth is often very difficult to accommodate, it is wiser to plan the complete installation around a residual current device, as shown in Figure 3.3. Regulation 412-06 refers.

Under no-fault conditions, the current flowing both in and out of the device is completely

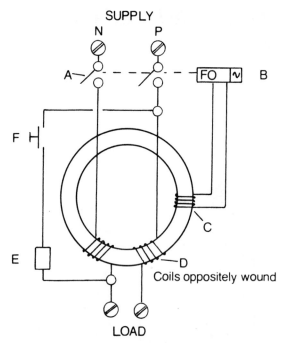

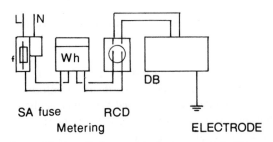

Figure 3.2. Residual current device (RCD): A double pole switch; B tripping relay; C fault-sensing coil; D toroidal transformer and induction coils; E test resistor; F test button.

Figure 3.3. A residual current device incorporated within an installation.

current to earth, the residual current device is electrically thrown out of balance. This allows a very small current to be induced into the fault-sensing coil and it is this output which provides the special polarised magnetic tripping relay with a signal, allowing the device to activate. This operates by cancelling the permanent magnetic flux by the excitation flux generated in the windings of the polarised magnetic relay. As this effect can occur only once every half cycle, the operation times can vary from 25 to 40 milliseconds (0.025 to 0.04 seconds at 50 hertz).

Sensitivities

The sensitivity of the apparatus is governed by the type of toroidal transformer and sensing relay used. Figure 3.4 outlines the basic requirements for a polarised or weakening trip relay.

Residual current devices are manufactured in a complete range of sensitivities from 2.5 to 1000 mA, but their working principles are often very different. Some work by electronically amplifying the output of the fault-sensing coil to operate a shunt trip. Others, as previously discussed, use a magnetic polarised tripping relay to directly release the mechanism responsible for mechanically activating the device. This technique is completely independent of the applied voltage but whichever type is used, protection is afforded by a combination of sensitivity and disconnection time. Generally an electronic amplified RCD is far slower than its polarised counterpart and has the added disadvantage that it is voltage-dependent.

balanced. This is achieved by the use of a ring-like transfomer called a *toroidal* in which the electrical load of the protected installation is fed through two opposing coils circumnavigating the laminated toroidal ring. In a healthy circuit the opposing coils would produce an equal and opposite magnetic flux and therefore no current would be induced into the supplementary sensing coil. When a fault condition occurs within the installation, either from a phase or neutral conductor leaking

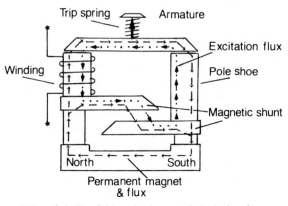

Figure 3.4. The field weakening or polarised trip relay.

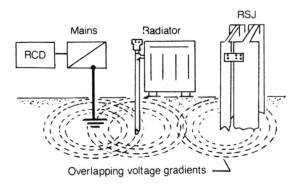

Figure 3.5. Examples of parallel earth paths.

Nuisance tripping

The residual current device is highly sophisticated in design and completely selective in operation when monitoring the circuit or installation it protects. Unlike the voltage-operated earth leakage circuit breaker, it is totally independent and unaffected by parallel earth paths which can cause nuisance tripping to occur. Figure 3.5 illustrates in graphic detail the way in which parallel earth paths can be inadvertently brought about.

From time to time the residual current device will trip for reasons other than a phase or neutral fault condition to earth. Ten of the more common causes why nuisance tripping will occur in an installation which is otherwise healthy are given:

1. Long runs of minerally insulated (MI) cable can act as a capacitor and activate the device. As an example, the nominal capacitance between conductor and sheath of 1 km of two-core, 1.5 mm^2 cable can reach as high as 0.22 μF.
2. Isolating or switching on rows of fluorescent lighting can often mislead the RCD to imagine that an earth fault has developed.
3. Isolating large electric motors can sometimes cause the device to act as though a fault condition to earth had occurred.
4. Spikes or anomalies in the supply voltage will initiate the trip.
5. Extraneous vibration may cause the residual current device to activate mechanically.
6. Lightning strikes targeted at the supply

authority's services cables generally cause nuisance problems unless the installation has been fitted with a special device with an intrinsic time delay.

7. Using the wrong type of radio frequency interference (RFI) filter will cause tripping to take place. It is worth remembering at this point always to check that the correct type of RFI filter has been used, as many are designed with capacitors connected directly to earth.
8. Large current surges lasting only milliseconds as in the case of a filament lamp rupturing may cause disruption within a residual current device.
9. Deteriorating cooker, immersion or night storage heating elements may cause nuisance tripping after a critical temperature has been reached.
10. A noisy residual current device often indicates an imminent fault or nuisance condition.

Should it be decided that a nuisance condition is unlikely and the residual current device still fails to latch in, either there is a fault to earth within the installation or an integral weakness has developed directly affecting the RCD.

To access either condition, the following procedure can be carried out:

1. Switch off the RCD.
2. Isolate all distribution fuse boards and loads which are being served by the RCD.
3. Disconnect all outgoing current-carrying conductors from the RCD.
4. Switch on the RCD.

If switching on causes the RCD to activate then it is the device which is faulty. In order to maintain a degree of safety within the installation, a new unit should be installed without delay. Never be tempted to bridge the switching movement in order to provide a service.

Once the RCD is installed, it is good practice to test regularly by pressing the test button, often marked 'T'. Residual current devices will not provide any protection against overcurrent; only current leakages to earth. They must always be installed in conjunction with a suitably rated fuse or circuit breaker distribution centre.

The voltage-operated earth leakage circuit breaker

Before the residual current device attained popularity, installations of special risk were protected by means of a *voltage-operated earth leakage circuit breaker* (ELCB). As with the RCD, voltage-operated earth leakage circuit breakers are mainly to be found in rural areas where local supply authorities provide a service by means of overhead cables.

Figure 3.6 clearly shows the internal wiring details of the circuit breaker and its relationship with the installation it is protecting. Known also as *earth-trips* or *fault voltage-operated devices*, they are principally designed to respond to a minimum of approximately 50 V entering the sensing coil in order to activate the switching movement.

In Britain it is no longer customary to install these devices because of the many disadvantages

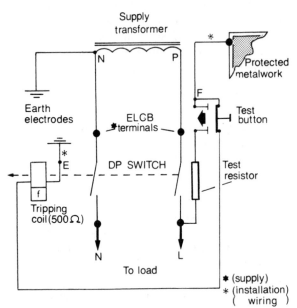

Figure 3.6. Voltage-operated earth leakage circuit breaker (ELCB). ELCB terminal E is connected to an earth electrode, whilst terminal F is connected by an insulated conductor to the principal earthing terminal and protective metalwork.

that accompany them. However, large numbers are still in service in country areas.

Practical principles
Working on the principle that should a live conductive part place a high potential on the surrounding exposed conductive metal work, a voltage-sensitive coil within the device will energise and activate a latching mechanism serving an integral switch, rendering the installation harmless.

The voltage-operated earth leakage circuit breaker is designed to function when a potential of 50 V has been reached within the faulty circuit. As with the residual current device, a test button is provided to test the sensitivity and mechanical operation of the apparatus. This should be tested regularly.

Disadvantages
One of the disadvantages is the unintentional shorting out of the voltage-sensitive coil. Unlike the residual current device, the ELCB requires an earth electrode to function properly. It is important to place the earth electrode outside the resistance area of any other extraneous metalwork which might be earthed. If this precaution is overlooked, it will have the effect of short circuiting the voltage-sensitive coil, so rendering the device inoperative. Figure 3.7 makes this clear. Another way in which the coil could be made ineffective is by use of an uninsulated copper conductor to serve the earth electrode from terminal 'E' in the device. Should the cable be placed in direct contact with earthed metal forming part of the installation or infrastructure it will short out the voltage-sensitive coil, making the apparatus unworkable.

Another weakness when using this type of circuit breaker to protect an installation is when two independent systems stake their relevant electrodes adjacent to each other. With their respective earth resistance areas overlapping, a fault condition in one installation can cause the ELCB in a neighbouring system to trip out as Figure 3.8 illustrates. A fault current crossing the overlapping earth resistance areas will find a conductive path to the tripping coil serving the fault-free installation.

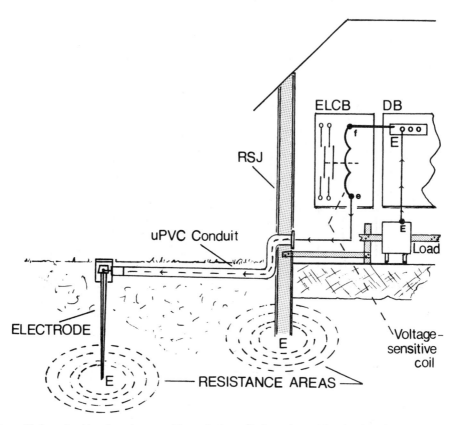

Figure 3.7. To avoid short circuiting the voltage-sensitive coil, the earth electrode must be placed outside the resistance area of any other extraneous metal which might be earthed.

At this point, any stray current will return through a convenient parallel earth path that is not within the resistance area of the electrodes and activate the healthy tripping device.

To remedy this weakness, both electrodes must be staked further away from each other so their respective areas are independent and do not overlap. Reference is made to Part 2 of the *Wiring Regulations*.

Figure 3.9 offers, by comparison, a typical wiring arrangement for a residual current device and voltage-operated earth leakage circuit breaker when incorporated into separate installations. Each system has its own earth electrode but for entirely different reasons. The earthing arrangement shown in the installation protected by a residual current device is to minimise any voltages occurring between exposed and extraneous conductive parts. The electrode serving the voltage-operated earth

leakage circuit breaker provides a return path for the voltage-sensitive element.

Fault-finding made easier

Phase or neutral fault conditions to earth can generate a great deal of unnecessary worry and frustration to an inexperienced operative. By knowing the correct procedure, much of the anxiety and unease may be eliminated.

Figure 3.10 shows a typical neutral to earth fault occurring within an installation serving an electric motor. During the initial wiring stage, the neutral conductor had been carelessly sprung against the metal-clad motor terminal housing. Constant vibration over the years had worn the insulation protecting the cable, thus causing a fault condition to arise.

To find the snag, a plan of action must first be

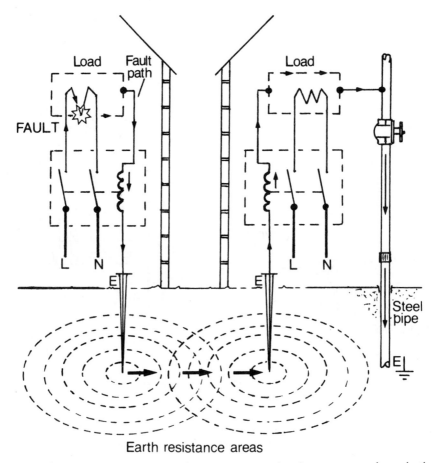

Figure 3.8. Nuisance tripping can occur when two or more earth resistance areas overlap each other.

chosen in order that logical steps may be taken leading to the fault. Ten stages are listed, some of which may be bypassed but usually experience will dictate how to proceed.

1. Switch off the installation. Remove main fuses.
2. Test the RCD by means of the test button. If satisfactory, proceed to item 3.
3. Remove all overcurrent protection and lamps from the installation.
4. Test all outgoing live conductors using a 500 V insulation tester (phase to earth).
5. If readings are satisfactory (half a megohm or more), proceed to item 6.
6. Disconnect all outgoing neutral cables and test to earth using a 500 V insulation tester.
7. Keep all neutral conductors in the same order as they were removed from the main neutral block.
8. The faulty cable will appear as a 'zero' or 'near to zero' on the insulation tester.
9. Identify the destination of the faulty cable through the neutral identification chart and follow it through to the fault.
10. Repair the fault. Reinstate the distribution board and lamps, then test again. The reading obtained should not be less than 0.5 MΩ. Regulation 713-04-04 confirms.

Low impedance faults
It is often an advantage to divide the faulty conductor at a convenient midway point in order to determine in which half the weakness lies. Faults

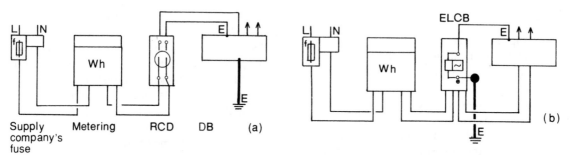

Figure 3.9. Incorporating current (a) or voltage-operated (b) circuit breakers into an installation.

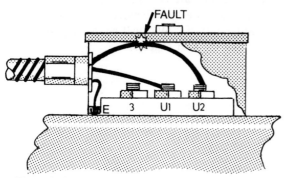

Figure 3.10. A common neutral to earth fault found in the terminal housing of an electric motor.

with very low impedances to earth may be made easier to find by using a digital insulation tester switched to the ohms scale. The closer the fault, the lower the meter reading will be. Should a general purpose multimeter be used, always return the selector switch to 'zero' or 'off' after use. This will avoid damage to the instrument if, inadvertantly, voltage is measured on the ohms scale!

Figure 3.11 illustrates the relationship between fault current and impedance when the supply voltage to earth is 230 V.

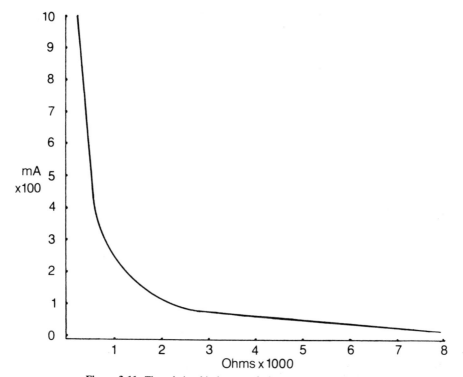

Figure 3.11. The relationship between fault current and impedance.

Summary

In this chapter we have considered the advantages and difficulties that can arise when installing current- and voltage-operated earth leakage circuit breakers.

We have seen how the two systems differ and how residual current devices, unlike voltage-operated, are completely unaffected by parallel earth paths, but are subject to nuisance tripping.

Logical means have been introduced by which faults may be identified by observing a plain, straightforward procedural pattern. This has been demonstrated by showing how a simple circuit serving an electric motor can develop a troublesome fault that will signal the residual current device to activate.

Because of their physical construction and often harsh environmental usage, electric motors served by a 'TT' distribution supply system seem to be far more prone to earth faults than other current-using appliances. Although this could be a debatable point, I have found this to be so especially when the motor forms part of an agricultural installation.

Electric motors are designed to provide a variety of services and are manufactured and moulded in many different ways. We shall be dealing with these within the next chapter.

4 Electric motors

In this chapter: Early concepts. Direct current and AC synchronous motors. Induction and commutator motors. Single phase and shaded pole motors. Basic motor control techniques. Power factor and correction.

Oersted, in 1820, was the first to discover and successfully demonstrate the effect brought about by a compass needle being placed near a current-carrying conductor. He was able to illustrate that the needle would always move at right angles to the conductor and would reverse direction when the current was changed. This fundamental principle was established long before *Faraday* had constructed his first workable dynamo in 1831.

Figure 4.1 illustrates the effect of placing a non-current-carrying conductor into a permanent magnetic field, whereas Figure 4.2 clearly displays the consequential effect and interaction when a current-carrying conductor is positioned within a similar field.

It has been shown how a load-bearing wire is completely surrounded by an invisible magnetic field in the form of ever-increasing concentric circles. It would, therefore, not be at all surprising to learn that the conductor would also experience a force upon it.

Placing two independent force fields within a

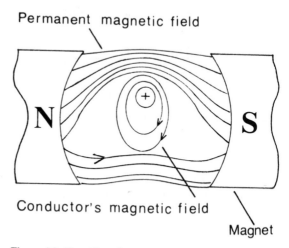

Figure 4.2. The effect of placing a current-carrying conductor into a permanent magnetic field.

magnetic environment would cause mutual interaction by spreading and buckling the lines of permanent flux into misshape. Imagine the lines of flux as elastic bands grossly distorted by the presence of a current-carrying conductor placed between the poles of a permanent magnet (Figure 4.2). The flux, acting in a similar manner to the elastic bands, will always try to adopt and maintain their original and natural, straight, parallel paths, and the only way this may be achieved is to push the conductor from the magnetic field. If, instead of a conductor, a loop is placed between the two opposing magnets, one side of the loop will be pushed up whilst the other will be forced down and a simple turning motion will be experienced. By increasing the number of turns on the coil and adding more coils to the original design, the effect of movement will be greatly heightened. This is the basic principle applied to the direct current (DC) electric motor. It is this we will be discussing next, but the detailed construction of the machine will not be debated.

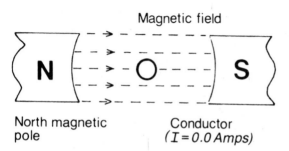

Figure 4.1. A non-current-carrying conductor is placed into a permanent magnetic field.

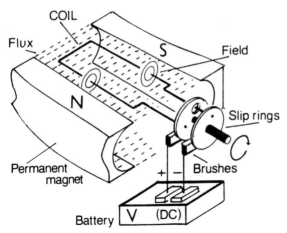

Figure 4.3. An undeveloped direct current electric motor.

Figure 4.3 illustrates the fundamental requirements for an undeveloped direct current electric motor.

Basic concepts

A coil of wire with slip rings attached to the open ends, illustrated as Figure 4.3, is sandwiched between two opposing poles of a permanent magnet. Should a DC voltage V be applied to the slip rings, in theory the coil will rotate. Whether the coil would actually rotate depends on the mechanical constraints of the armature coil in relation to the pole pieces. A rotating coil, cutting through the invisible lines of magnetic flux as illustrated, will induce a secondary electromotive force E into the coil. At this point the device is acting not only as a simple electric motor but also as a generator.

We have seen in the previous chapter that a generated electromotive force will *oppose* the applied voltage V, and it is this induced voltage E that is sometimes referred to at the *back emf*.

Calculating the back emf

By measuring the current taken from the motor and evaluating the voltage applied, the back emf may be calculated by use of the following expression:

$$I_a = \frac{V - E}{R_a} \qquad [4.1]$$

where I_a is the current in amps generated in the coil or armature,

V is the applied voltage,

E is the back emf in volts and

R_a is the resistance of the coil or armature in ohms.

First Expression [4.1] must be arranged in terms of E by cross-multiplying:

$$I_a \times R_a = V - E \qquad [4.2]$$
$$\therefore (I_a \times R_a) + E = V \qquad [4.3]$$
$$E \text{ (back emf)} = V - (I_a \times R_a) \qquad [4.4]$$

As a practical example, consider the following problem:

A small 220 V electric DC motor has a measured coil resistance of 80 Ω and found to be drawing a current of 2.6 A from the supply. Calculate the back emf produced in the motor.

By referring to Expression [4.4]:

$$E = V - (I_a \times R_a)$$

Substituting figures:

$$E = 220 - (80 \times 2.6)$$
$$E = 220 - 208$$
$$E = 12 \text{ volts, back emf}$$

This is very much an academic example but it will help to assist and clarify how a back emf may be calculated.

Basic DC motors and starting controls

Starting a large DC electric motor can lead to problems coupled to current control, so in practice a current-limiting device is incorporated into the circuit. Figure 4.4 shows such a device which has been designed to incorporate an overload-sensing facility controlling the function of a 'no volt coil'. The illustration depicts the role a DC starter plays when serving a typical direct current shunt motor. Increasing the resistance will decrease the magnetic flux and so increase the speed of the motor. In practice the permanent magnets referred to at the beginning of this chapter are replaced by electromagnets called field coils. These, in a *shunt*

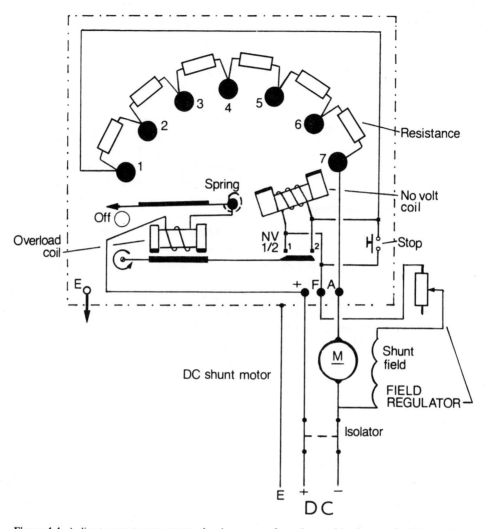

Figure 4.4. A direct current motor starter showing means of speed control by the use of a field regulator.

wound motor, are wired in parallel with the armature, whereas in a *series wound* motor, as depicted in Figure 4.5, the field coils are wired, as the name suggests, in series with the armature. The current flowing in the armature will always be the same as the current circulating throughout the field coil.

Series wound motors may be used either on AC or DC supplies. They have a good starting torque but there is a notable drop in armature speed when load is applied. Removing the load will lead to the motor racing to extremely high speeds, so this class of motor should not be employed where constant speed is required. Series motors are

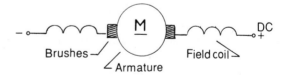

Figure 4.5. A direct current series wound motor.

usually rated below 0.2 kW, and are mainly used in portable hand tools and small electrical appliances found throughout the home. Another type of motor applied to DC installations is the *compound motor*, which as the name suggests, is a combination of both series and shunt wound, as shown in the schematic arrangement of Figure 4.6.

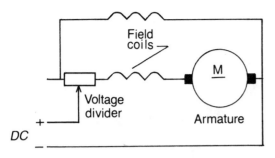

Figure 4.6. DC compound wound motor.

Alternating current motors: three phase

There are so many different types and varieties of motors available today, it would necessitate a book of this size in order to examine and investigate the theoretical concepts and to study the basic characteristics and applications for each type of machine listed. The aim of this chapter is to explain theoretical concepts involved behind the various classes of AC motors. These may be arranged into several individual groups, the most common being:

1. Synchronous motors
2. Induction motors
3. Commutator motors

Figure 4.7 illustrates how the AC motor is divided into three main identifiable parts. These comprise:

1. Stator — the stationary member carrying the field windings and connected to the supply.
2. Rotor — cylindrical core mounted on a shaft, free to rotate within the stator.
3. End shields — moulded circular discs accurately positioned on the ends of the

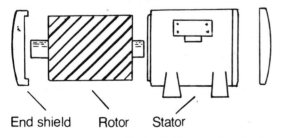

End shield   Rotor   Stator

Figure 4.7. The three main identifiable parts of an electric motor.

stator to support the shaft within suitable bearings.

The synchronous motor

Consider first a simple synchronous motor in its most basic form relying entirely on a rotating magnetic field generated within its stator windings. The speed of the rotor is determined by the number of pairs of stator poles and the frequency of the supply voltage. Figure 4.8 shows how the rotor adopts the form of a permanent magnet chasing a rotating magnetic field generated within the stator windings.

Simple synchronous motors are not self-starting but have to be rotated to the speed of the electromagnetic field. Once this has been achieved, the motor will 'lock-in' and pace the field to maintain a constant speed. Should the applied load become too great, the rotor will find difficulty in pacing the gyrating magnetic field and will stall. This type of motor may be designed for use on either single or three phase supplies. An example may be drawn from the domestic mains voltage clock or an industrial time-switch employed to control current-using appliances automatically. Larger industrial synchronous machines, often used to run air conditioning plants, will be discussed more fully in Chapter 13.

The induction motor

Induction motors are the most common of all types of motors used today and may be subdivided into

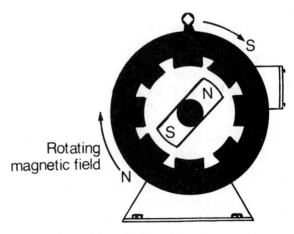

Rotating magnetic field

Figure 4.8. A simple synchronous motor.

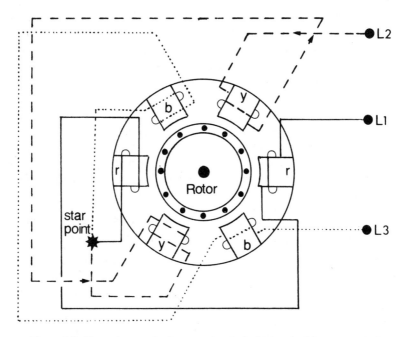

Figure 4.9. Three phase squirrel cage motor: typical internal wiring arrangements.

two classes known as *squirrel cage* and *double cage*.

As with the synchronous machine, the squirrel cage motor may be selected for use on single or three phase supplies. Figure 4.9 illustrates how the windings are situated within the inner periphery of the stator, and their relationship to the supply voltage. These are recessed into laminated soft iron slots and accommodated in a rolled steel housing.

When a three phase supply is applied to the stator windings, a rotating magnetic field is produced. However, the rotor is in no way connected to the supply voltage but takes the form of a cylindrical soft iron laminated component. This is mounted on a shaft running through the centre of the rotor, allowing it to spin unhindered within the stator. The shaft is supported by bearings housed in end plates.

Copper bars are drawn into hollowed slots which run at intervals throughout the length of the rotor and are made common with each other by means of a brass ring brazed to each end of the copper bars. Figure 4.10 shows how the short circuiting rings are attached to the copper bars.

This class of machine is known as *squirrel cage*.

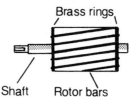

Figure 4.10. The squirrel cage rotor (see also Figure 13.34).

The rotor windings are formed in a manner resembling the type of cage in which pet squirrels were once exercised.

In Chapter 1, electromotive force and current flow were reviewed and it was shown how a coil of wire cutting through a magnetic flux at right angles to it would induce an emf in the coil. It is this same electrical principle that enables a squirrel cage motor to function by working to the precept that a rotating magnetic field produced in the stator coils induces an emf into the rotor winding by 'transformer action', often referred to as mutual induction.

Since the windings are short circuited by brass rings positioned at either end of the rotor, currents will be able to flow throughout the rotor circuit. They too will set up their own magnetic fields and in doing so will interact with the field created within the stator windings. The stator's rotating magnetic field, known as the *synchronous speed* of the motor, has the effect of dragging the rotor around with it. The current produced in the closed loop of copper conductors serving the rotor opposes the very current producing this phenomenon. The only way to counter this change is for the closed loop, the rotor, to gyrate in the same direction as the rotating magnetic field generated within the windings of the stator. In practice the rotor will follow at a little less than the synchronous speed. If both speeds were matched, the rotating magnetic flux would be unable to cut through the rotor windings. An emf would not then be induced into the copper bars serving the rotor to produce the electromagnetic field that would enable it to function.

Slip The difference between the synchronous and the rotor speed is known as the 'slip' and can vary from approximately 2.5 to 5.5 per cent at normal loads. If the load is greatly increased, the difference between the two speeds will also increase and, if overloaded excessively, the motor will stop.

Slip is usually expressed as a percentage of the synchronous speed of the motor and can be calculated by employing the following expression:

$$\text{slip } \% = \frac{N - \text{rotor speed}}{N} \times \frac{100}{1} \quad [4.5]$$

where N is the synchronous speed of the motor.

As an example consider the following:

An induction motor has a synchronous speed of 1500 revolutions per minute (rpm) when connected to a 50 Hz supply. If the rotor speed is 1450 rpm, the percentage slip would be:

Referring to expression [4.5] and substituting figures:

$$\text{percentage slip} = \frac{1500 - 1450}{1500} \times \frac{100}{1}$$

$$\text{percentage slip} = \frac{50 \times 100}{1500}$$

$$\therefore \text{ percentage slip} = 3.33\%$$

In practical terms, if the synchronous speed of the

motor is unknown, it may be calculated, providing prior knowledge as to the number of pairs of poles has been gained in advance, by use of the expression:

$$N = \frac{f}{p} \quad [4.6]$$

where N is the synchronous speed of the motor in revolutions per second,
f is the frequency of the supply in hertz per second, and p is the number of *pairs* of poles.

Expression [4.6] may be interpreted and made clearer by advancing the following problem:

Calculate in revolutions per second (rps) the rotor speed of a four pole, 50 Hz induction motor which has a previously calculated 4 per cent slip on full load.

Referring to Expression [4.6]:

synchronous speed

$$= \frac{\text{frequency in hertz}}{\text{number of pairs of poles}}$$

Substituting for known values (remember — four poles but two pairs):

$$N = \frac{50}{2}$$

Theoretical speed $N = 25$ rps

Calculating the true speed with a known slip of 4 per cent:

$$0.04 \times 25 \text{ (i.e. 4\% of 25)} = 1 \text{ rps}$$
$$\text{actual rotor speed} = 25 - 1$$
$$= 24 \text{ rps}$$

It is common to express the speed of a motor in revolutions per minute and this may be achieved by multiplying by 60:

$$\therefore 24 \text{ rps} \times 60 = 1440 \text{ revolutions per minute}$$

Squirrel cage induction motor In the United Kingdom the supply frequency is 50 Hz. By calculation it can be seen that the range of speeds in revolutions per second is surprisingly limited. Take as an example a machine with a single pair

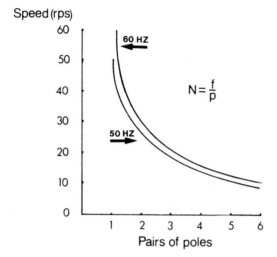

Figure 4.11. The relationship between frequency, pairs of poles and speed.

$$N = \frac{f}{p}$$

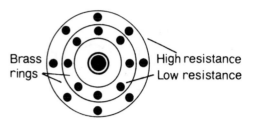

Figure 4.12. End view section of a double cage induction motor rotor.

of poles. By applying Expression [4.6] it can be calculated that the machine will deliver a maximum theoretical speed of 50 rps or 3000 rpm. Two pairs of poles will reduce the machine's speed to 25 rps or 1500 rpm. Three pairs of poles would allow the synchronous speed to fall to 16.6 rps and so on. A higher supply frequency will produce a higher synchronous speed as Expression [4.6] and Figure 4.11 clearly demonstrate.

The squirrel cage induction motor is one of the most popular types of motor used today. Not only is it durable and hardwearing but also reasonably priced and extremely efficient. The machine is easy to maintain and the control equipment is readily obtainable. Unfortunately there are few practical speeds that can be obtained, as Expression [4.6] illustrates. Other disadvantages are that the starting torque is low and the current can be up to three or four times greater than the running current. The induction motor may be employed in a variety of applications. Many are found in industry and in agricultural installations, serving fans, conveyors, feed augers and vacuum pumps where a constant speed is essential.

Double cage induction motor A close cousin to the squirrel cage induction motor is the *double cage induction motor* which has been designed and engineered in a similar fashion. Instead of just one set of rotor bars there are two, as Figure 4.12

shows. The lower cage has a higher resistance than the upper cage. When current circulates through both sets of rotor bars it produces a far more effective starting torque, enabling a lower starting current to be achieved. The lower resistance bars act in a similar way to the standard squirrel cage machine, producing a high efficiency factor on load.

The main disadvantage of this motor lies in its cost, but where high starting torque is required, as with water pumps, air and refrigeration compressors, this type of motor will always be found.

Commutator motors
The triple phase *commutator motor* is extremely expensive and complicated in construction. Brushes, serving the commutator, are designed to alter their position automatically in order to produce a wide range of working speeds. This type of motor is generally used in place of a DC machine where it is essential to maintain a high degree of speed control as, for example, a printing press. In the nature of its speciality, the theoretical concepts and characteristics of this machine will not be discussed but the need to introduce this type of motor is relevant.

Slip ring motor As with the triple phase commutator motor, the *slip ring motor* can be described as equally expensive and complicated. It could well be portrayed as hybrid since it has both wound rotor and stator wired in star formation. Figure 4.13 illustrates how the ends of the rotor windings are connected to three brass slip rings which are mounted, but insulated from, a central steel shaft forming the horizontal axis of the rotor. A comparison can be made at this stage between

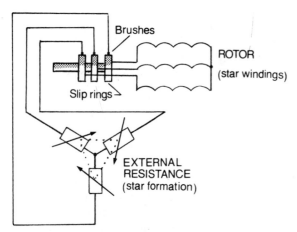

Figure 4.13. The windings of a wound rotor induction motor are connected to three brass slip rings mounted, but insulated, on the steel shaft forming the axis of the rotor.

this and the squirrel cage induction motor, as there is no physical electrical link between the stator and the rotor. Current flowing in the rotor windings is the direct consequence of mutual induction. The supply voltage is connected to the field coils of the stator.

The three slip rings are served by spring mounted carbon or copper brushes leading to a bank of variable resistors wired in star connection. These can be automatically or manually programmed to cut in and out of the rotor circuit, adjusting the current to achieve the required speed control.

It is usual to start slip ring machines with the maximum resistance in circuit and then gradually to reduce resistance as the motor gains speed. At full speed all external resistance is removed from the rotor circuit and the rotor is effectively short circuited. When this stage is reached the motor will act as a typical squirrel cage induction motor.

By varying the value of the rotor control a wide choice of torque and speed is obtainable. An added advantage of this versatile machine is that the starting current may be regulated. The disadvantages are mainly financial and also that complicated control gear requires regular maintenance.

Heat losses developing in the rotor control can decrease the efficiency of the motor and weaken the effect of speed control on load. Although

variable speed is attainable, the cost is a reduced motor efficiency.

AC motors: single phase induction motor

Previous paragraphs have shown that the three phase induction motor is completely dependent on a rotating magnetic field generated within the windings of the stator.

Single phase induction motors are not inherently self-starting, and function subject to a pulsating magnetic field. Consequently, the torque produced is also pulsating and not constant, but if the rotor is spun, will accelerate until a speed is reached which is just below that of the synchronous speed. This is much the same way as the triple phase induction motor operates.

This type of machine can be tailored to be self-starting by adding another winding, displaced exactly 90 degrees electrically on the stator and connected in parallel with the original winding. Known as a *split phase induction motor* or *resistance start induction motor* the two windings, wired in tandem, produce a two phase rotating magnetic field due to a difference of impedance. In practice, the 'start' winding has a higher resistance then the 'run' winding and is taken out of circuit automatically by means of a centrifugal switch or timed relay. This is illustrated in Figure 4.14. As the start and run windings are not pure inductors but have an element of resistance in them owing to the natural resistance of the coils, it causes the

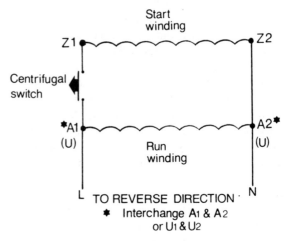

Figure 4.14. Split phase induction motor.

current flow to be less than 90 degrees out of phase with the voltage. In practice this type of motor will develop a phase differential of about 40 degrees between the current in the start windings and the current in the run windings. This results in poor starting torque and a slight drop in speed when load is applied.

The split phase induction motor is employed when the starting load is small or there is no load at all, as, for example, in larger domestic appliances such as vacuum cleaners, or workshop pedestal drills.

Capacitor start and capacitor start and run motors

To improve torque and phase differential in single phase induction motors the merits and characteristics of the *capacitor start* and *capacitor start and run motor* will be examined. Figures 4.15 and 4.16 illustrate these two types of motor in basic schematic detail.

Capacitor start motor

By placing an electrolytic capacitor in series with the start winding, a greater phase angle and starting torque may be achieved. The start winding, now capacitative, motivates the current to lead the voltage to achieve a greater phase differential between the two sets of windings. By adding the correct value capacitor, a phase angle

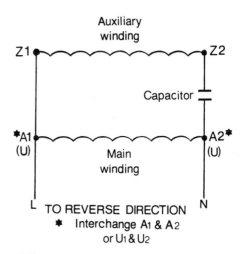

Figure 4.16. Induction split phase (capacitor-aided) motor. Also known as capacitor start and run motor.

of 90 degrees can be developed between the two sets of currents and therefore maximum starting torque will be achieved.

As with the split phase induction motor, the start winding is automatically isolated by means of a centrifugal switch, or timed relay, which is activated when approximately 75 per cent of the maximum rotor speed has been reached.

Many manufacturers do not recommend starting this class of motor any more than seven or eight times per hour for fear of damaging the capacitor. In practice, capacitor start machines are usually fitted with capacitors within the 240 volt working range, whereas capacitor start and run motors utilise those with a much higher working voltage.

Split phase capacitor start induction motors are designed to produce high starting torque but experience a slight drop in speed when load is applied. They are often used to serve compressors and pumps where high initial starting torque is required.

Capacitor start and run (capacitor-aided) motor

A close relation to the capacitor start, sometimes called the *permanent-split capacitor motor*, is the *capacitor-aided motor*. As the name suggests, the capacitor is kept in circuit continually and assists in correcting the power factor as Figure 4.17 shows. (See also under *Power factor correction*, towards the end of this chapter.) This eliminates

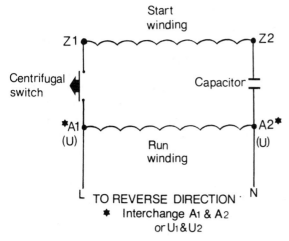

Figure 4.15. Induction split phase (capacitor start) motor.

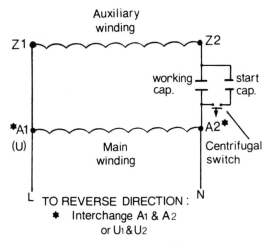

Figure 4.17. Capacitor-aided motor.

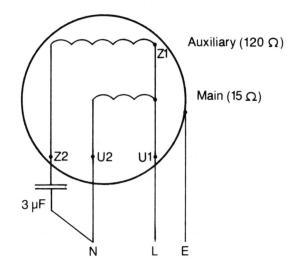

Figure 4.18. Typical internal wiring arrangement serving a capacitor-aided motor. Terminals U1 and U2 supplying the main winding were once known as A1 and A2.

any need to employ a centrifugal switch or timed relay. Generally, oil-filled 400 volt working voltage capacitors are employed to serve this class of motor. Often a second capacitor is added to the circuit and placed in parallel with the run capacitor to aid starting. When this technique is used, the 240 volt auxiliary capacitor is taken out of circuit by means of a centrifugal switch or timed relay. This is designed to function when approximately 70 per cent of the maximum potential speed has been reached.

The rotor used is of typical squirrel cage construction as described above.

General characteristics The capacitor-aided motor has very good starting and running torque and may be often found in the air conditioning field used within refrigeration plants or serving industrial unit heaters. A slight drop in speed can be experienced with load.

On-site wiring A great deal of unnecessary frustration may be avoided by understanding the correct method of wiring a motor of this type. Some machines are prewired in such a manner that it is very easy to understand how both motor and capacitor may be connected to the electrical supply. However, the method of connection is not always so obvious with other types of machines, especially those employed to serve refrigeration condenser units. This is because the capacitor

serving the motor is detached and separately housed in a remote enclosure well away from the motor.

This class of motor has three conductors and a current-protective cable originating from its stator. Most conductors are separately identified by the use of colours but sometimes only mono-colours are used if the machine has been rewound. There are obviously several different wiring combinations that could be tried in order to connect the capacitor and motor to the mains supply, but only one way is correct.

Figure 4.18 illustrates a typical capacitor-aided motor employed to serve a fan for use in a refrigeration plant. The motor comprises two sets of windings: one higher in resistance than the other but both sharing a common conductor. A capacitor of the correct value and working voltage is connected in series formation with the motor winding registering the highest impedance. The unterminated side of the capacitor can then be connected to the incoming phase conductor. The supply neutral will then be wired to the unconnected tail of the highest value winding. The lower valued winding is also connected to the incoming supply, but is fed directly.

Wiring this way removes the guesswork and makes life a little easier!

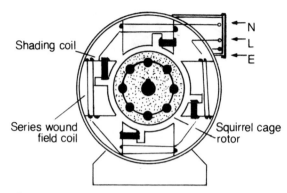

Figure 4.19. Four pole shaded pole motor.

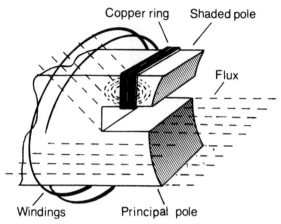

Figure 4.20. Shaded and principal pole pieces.

The shaded pole induction motor

During the past few years the shaded pole
induction motor has become more widely accepted,
mainly due to its new-found reliability coupled
with a simple trouble-free design.

The machine comprises a wound stator and
squirrel caged rotor, (Figure 4.19), and it might at
first sight be confused with a resistance start
induction motor. This is because neither start
winding, capacitor nor centrifugal switch is
included within the design.

The speed is determined by the same criteria and
electrical factors governing other types of squirrel
cage motors and operates by reversing the
magnetic polarity between the principal pole piece
and the shaded pole. The pole, placed adjacent,
forms an integral part of the principal magnetic
pole and is served with a low resistance copper
band around its periphery.

The rapid change of magnetic polarity
experienced produces a rotating magnetic field
and, as with other types of induction motors,
induces an emf into the rotor windings. Since the
conductors serving the rotor are short circuited to
form a squirrel cage, they too develop their own
magnetic field interacting with the field produced
by the stator windings. As the stator produces a
rotating magnetic field it has the effect of dragging
the rotor around with it at a speed slightly less
than the synchronous speed. Figure 4.20 illustrates
the construction of the complete pole piece and
will help clarify the relationship of the two poles
to each other.

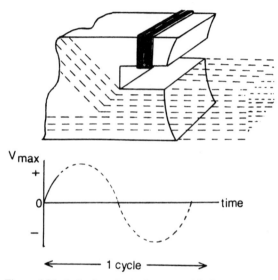

Figure 4.21. A rise in voltage allows magnetic flux generated
by the field coil to cut through the shading pole, inducing an
electromotive force in it.

Theoretical concepts To understand the theoretical
concepts of the shaded pole induction motor, first
imagine the field coil as the primary winding of a
transformer; and the copper band, the shading
pole, as the secondary winding.

As the voltage wave rises in magnitude from
zero to a near maximum voltage, illustrated as
Figure 4.21, magnetic flux generated by the field
coil cuts through the solid copper band placed
around the shading pole, inducing an emf in it.

The low resistance of the band produces a high
current and a magnetic field which opposes the

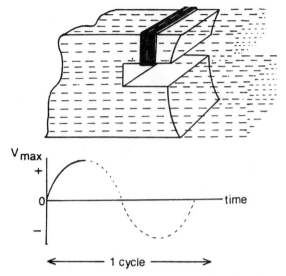

Figure 4.22. The flux is completely uniform throughout the pole piece when the voltage is not changing.

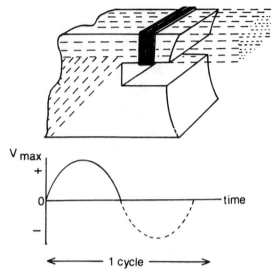

Figure 4.23. A decrease in voltage allows the shading coil to oppose a change of flux and the process starts again once the voltage passes from zero.

rapid change in flux in the principal pole until a peak voltage has been reached (Figure 4.21). At this stage the flux is completely uniform throughout the pole piece and the voltage will start to decrease and head towards zero again (Figure 4.22). The rapidly diminishing voltage causes the flux to collapse and induces an emf in the shading coil. Flux formulated by the induced emf opposes the change of magnetic flux in the principal pole, causing it to be concentrated in the shaded pole as Figure 4.23 illustrates.

When the voltage passes zero and travels towards the negative half cycle (Figure 4.23), the complete process starts again. This time the magnetic fields are reversed and it is this change once every half cycle that causes a rotating field across the pole piece.

Reversing the rotation of the machine is not at all practical as the stator windings would have to be completely removed, turned around and reinstated. This class of motor is usually classed as non-reversible.

Use and general characteristics The shaded pole induction motor may be found in air conditioning units serving fans and blowers and is widely used in the recorded music industry. A slight drop in speed is experienced when load is applied but

torque is dependent on the strength of the magnetic field and the phase angle differential between the current in the stator and the current flowing in the squirrel cage rotor.

Direct-on-line control gear for single and three phase induction motors

Control gear serving popular ranges of single and three phase squirrel cage induction motors is both reasonably priced and easy to maintain. The majority of starters in use today will accommodate a range of voltages; only the coil within the device need be changed if necessary. Figure 4.24 outlines the basic internal wiring to be found in a typical direct-on-line starter.

Pressing the 'start' button will immediately bring the control circuit into play by energising the coil and activating the load switching assembly to provide a direct supply to the motor. When energised, auxiliary contacts 7 and 8 maintain the control circuit in operation until such time as the circuit is interrupted by use of the 'stop' button or is automatically tripped because of overcurrent. This effectively isolates the coil circuit between terminals 95 and 96.

The thermal overloads will only activate upon a sustained increase of current flow and will then

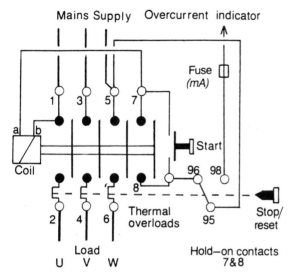

Figure 4.24. A typical direct-on-line three phase starter.

automatically close down the control circuit by triggering the tripping movement, wired in circuit between terminals 95 and 96.

A remote warning indicator can, if required, be wired from terminal 98 in order to register an overload condition. This should be accompanied with a suitably sized overcurrent protection device (Regulation 435-01-01).

Mechanical starters

An older and less favoured type of direct-on-line starter is the *mechanically activated starter*, designed and constructed for mechanical switching of smaller loads of up to 2.5 kW. This form of control functions in a purely mechanical manner and so does not employ an integral control circuit to serve a coil, which when energised would bring a switching cradle into play.

Overcurrent protection is affected by means of thermal overloads, taking the form of bimetalic strips attached to mechanical tripping levers enabling the supply to be isolated from the machine. Usually overcurrent protection is fixed, offering only two working current ratings to serve the circuit.

The mechanical starter has decreased in popularity over recent years but there are still many to be seen, especially within agricultural or light engineering situations. The starter has the advantage of being relatively inexpensive and easy to wire. Protection afforded to the motor and circuit is not so accurate as its modern counterpart. In addition, the mechanical components of the device require occasional maintenance in order to sustain reliability throughout their working life. The direct-on-line mechanical starter may be used on either single or three phase motor circuits of up to 2.5 kW.

Power factor correction

The power factor of an electrical circuit is the ratio of the true power in watts dissipated, to the product of the voltage and current and is expressed as a number. Hence,

$$\text{power factor} = \frac{W}{VA} \qquad [4.7]$$

where W is the power in watts and
VA is the product of the voltage multiplied by the current.

It is in every supply authority's interest to provide generated power to a consumer whose installation power factor is unity. Only when this is achieved can all potential power generated be effectively used. Should this concept be taken to extremes, for example in the case of a customer's power factor being zero, the supply authority's alternator would be working at a maximum efficiency but no power could develop within the system.

Consider a small private 230 V alternator capable of delivering a maximum output of 100 A (23 kW), supplying a dairy farm whose installation power factor is 0.55 or 55 per cent.

Since the power factor is 0.55, the herdsman will only be able to consume 0.55 of 23 kW or 12.65 kW. Obviously it is in the owner's commercial interest to encourage the customer to increase the value of the installation power factor to nearer unity, as he will only receive 55 per cent of any potential revenue. A factor of unity will provide an extra 45 per cent to be drawn from the alternator, to enable installation expansion or to serve another customer.

The vast majority of domestic circuits have a

power factor near to unity, but industrial, commercial and agricultural installations employing inductive circuits, such as motors and fluorescent lighting, often have a power factor far less than is desired.

Bringing the power factor to unity

The value of a power factor may be increased by connecting a suitable sized capacitor in parallel with the inductive circuit under review. Figure 4.25 illustrates an alternator serving a simple circuit forming a capacitor, resistor and inductor connected in parallel formation. When the capacitative current is at its peak positively, the inductive current is at its peak negatively, and it is this phenomenon that can be put to use to cancel one another (Figure 4.26). The current flowing in the capacitative circuit is directly opposed to the current flowing in the inductive circuit; thus they are 180° out of phase. Putting it another way: current flowing in the capacitative circuit leads the

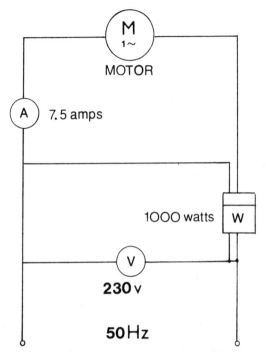

Figure 4.27. Evaluating power, current and voltage by instrumentation for the purpose of calculating the power factor correction.

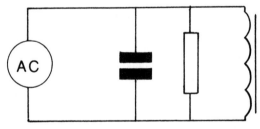

Figure 4.25. A parallel circuit forming a capacitative, resistive and inductive load.

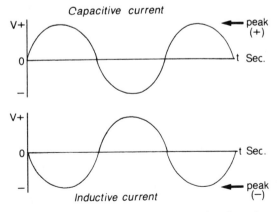

Figure 4.26. When the capacitative current is at its peak positively, the inductive current is at its peak negatively.

voltage by 90 degrees, whereas the inductive current lags behind the voltage by 90 degrees. As both capacitative and inductive currents are cancelled out, the alternator responds to only the resistive circuit.

As a practical example and an aid to understanding the concept of power factor and correction, consider the following problem:

A 230 V induction motor is metered to establish the value of both power and current flowing in the circuit for the purposes of power factor correction and evaluation. Figure 4.27 illustrates that the power dissipated within the circuit is 1000 W and that a measured current flow of 7.5 A has been recorded.

Calculate the value of the capacitor required to bring the power factor to unity.

By referring to Expression [2.4] the apparent power consumed in the circuit may be calculated:

$$\text{power in watts (volt-amps, } VA) = E \times I$$

Substituting for figures,

$VA = 230 \times 7.5$

$VA = 1725$ watts

The power factor can now be calculated by use of Expression [4.7]:

$$\text{power factor} = \frac{W}{VA}$$

Substituting for known values,

$$\text{power factor} = \frac{1000}{230 \times 7.5} = 0.57$$

This may be transferred in terms of a percentage by multiplying by 100. Hence,

$$100 \times 0.57 = 57\%$$

To enable this extremely low power factor to be corrected and brought to unity, we must first determine how much of the circuit is reactive power (volt-amps) and how much is true power (watts).

To recap: the value of the true power has been instrumentally measured and shown to be 1000 watts. The apparent power, VA, has been calculated and found to be 1725 watts. By use of the figures available, the reactive power may now be calculated by means of the following expression:

$$\text{reactive power} = \sqrt{(VA^2 - W^2)} \qquad [4.8]$$

Substituting figures,

$$\text{reactive power} = \sqrt{(1725^2 - 1000^2)}$$
$$\text{reactive power} = \sqrt{1\,975\,625}$$
$$\text{reactive power} = 1405.56 \text{ watts}$$

Since the electric motor is an inductive device, the reactive power generated may be cancelled out by an equal amount of capacitative reactance in the form of a suitably sized capacitor wired in parallel with the motor supply terminals.

The value of capacitative reactance required to increase the power factor to unity may be found by the following expression:

$$X_C = \frac{E^2}{\text{reactive power}} \qquad [4.9]$$

where X_C is the capacitative reactance in ohms and
E is the applied voltage.

Enumerating,

$$X_C = \frac{230^2}{1405.56}$$

$$X_C = 37.63 \text{ ohms}$$

By using Expression [4.10], the amount of capacitance required to produce 37.63 ohms of capacitative reactance can be calculated:

$$X_C = \frac{1}{2\pi f c} \qquad [4.10]$$

where f is the frequency of the supply voltage in hertz,
c is the capacitance in farads and
π is valued at 3.1416.

By cross-multiplying and dividing each side of the equation by $2\pi X_c f$,

$$C = \frac{1}{2\pi f X_C} \qquad [4.11]$$

Substituting figures,

$$C = \frac{1}{2 \times 3.1416 \times 50 \times 37.63}$$

$$C = 0.0000845 \text{ farads}$$

This figure may be converted to microfarads (μF) by multiplying by 1 000 000. Hence

$$1\,000\,000 \times 0.0000845 = 84.5 \ \mu F$$

A capacitor connected in parallel formation with the supply voltage serving the motor will effectively increase the power factor from 0.57 to unity. In practice it is difficult to achieve a unity power factor as the exact valued capacitor can seldom be met (Figure 4.28).

Summary

In this chapter simple direct current motor principles were examined. This was followed by basic theoretical concepts giving prominence to the series, shunt and compound motor. Speed control was shown to be effective by use of a variable linear resistance connected in series formation with the armature and field windings and operated manually. Basic principles of the alternating

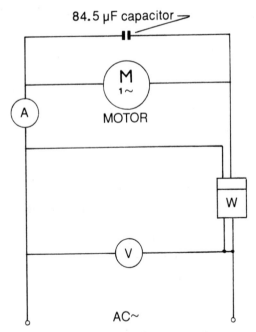

84.5 µF capacitor

A

M
1∼

MOTOR

W

V

AC∼

Figure 4.28. Power factor corrected.

current motor were discussed, highlighting their total reliance on either a pulsating or rotating magnetic field to act as a prime mover.

Calculations, establishing the speed of an induction motor, have been made by use of Expression [4.6] ($N = f/p$). 'Slip' was shown to be expressed as a percentage of the theoretical speed and calculated by use of Expression [4.5] ($N -$ rotor speed $\times$ 100/N).

By connecting a suitable capacitor in series with the start winding, it was demonstrated that a far greater phase angle and starting torque could be achieved.

Capacitors employed for use with AC machines were found to be extremely versatile, acting either as a continuous aid or as a means for automatic starting. A minority were connected in circuit to provide an improved power factor for the installation.

In Chapter 5 we will be exploring the basic theoretical boundaries of the capacitor, inductor and resistor. These three fundamental electrical quantities form the corner-stones for higher concepts in theoretical electrotechnology.

5 Resistance, capacitance and inductance

In this chapter: Simple capacitors. Capacitors and alternating current. Series and parallel formation calculations. Practical applications. Circuits containing capacitance, resistance and inductance. Capacitors — types and usage.

In Chapter 4 theoretical principles governing the performance of the single phase induction motor were considered. It was demonstrated that capacitance afforded an essential ingredient in enabling initial starting. Wired in series formation with the auxiliary windings of an induction split phase machine, the capacitor was shown to provide a means for continuous running.

The simplest form of capacitor can be constructed by placing two similarly shaped metal plates squarely together but separated with a special type of insulating material called a *dielectric* as illustrated in Figure 5.1. The total capacitance, measured in farads or subdivisions of farads, is determined by three factors which, after calculation, are built into the component at the design stage. The *surface area*, the *distance between the plates* and the *type of dielectric used* are the three main criteria used to regulate the final value of the capacitor.

Dielectric strength

Dielectrics are produced using many different materials. However, it is vitally important that the voltage applied to the plates is not responsible for breaking down the insulating dielectric separating them. Should this happen, the capacitor will be destroyed, so it is essential to choose the correct type and value of dielectric to suit the design of the circuit.

The voltage per unit thickness of the dielectric is the principal value that determines the working efficiency of the capacitor. This is referred to as the *dielectric strength* and is the maximum electric field strength the insulating material is able to withstand before breaking down through heat.

Listed below is a selection of examples in general use. Values are in kilovolts per millimetre thickness:

Mica (used for high frequencies): 40–200
Waxed paper (used for low frequencies): 40–60
Glass (used in high frequencies): 5–30
Paper (used for low frequencies): 4–10
Air (used for radio tuning): 3–6

Figure 5.2 depicts the terminals of a 12 V battery connected to a simple capacitor. Chapter 1 has shown that electron flow is a stream of negatively charged sub-atomic particles. As long as

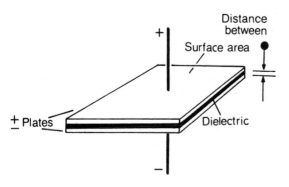

Figure 5.1. A simple capacitor.

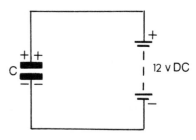

Figure 5.2. A 12 V battery connected to a simple capacitor (C).

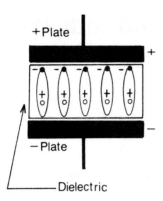

Figure 5.3. The orbits of the electrons become elliptically deformed. This is known as dielectric stress.

an electrical pressure is applied across the two plates, the positive plate will constantly attract an electron flow from the negative plate. This will continue until the voltage across the capacitor is equal to that of the battery. Disconnecting will leave the capacitor in a fully charged condition. At this stage, negative electrons serving the material forming the dielectric are repelled from the fully charged negative plate and attracted to the positive plate. Remember, as with magnets, *like poles repel* each other whilst *unlike attract*. The electrical pressure between the two opposing plates produces an elliptical deformity within the orbits of the electrons, forming the sandwiched dielectric as illustrated in Figure 5.3. It is this atomic deformity which is known as *dielectric stress* and is directly proportional to the voltage applied.

Should the voltage be increased to a point beyond that of the dielectric strength, a physical breakdown will occur within the capacitor, causing an electrical leakage from one plate to another. In practical terms this often takes the form of the dielectric burning through and shorting out the polarised plates. Once this has happened the device is completely unserviceable and should be replaced.

It is worth remembering at this point that capacitor aided motors are generally attended by capacitors having a working range of 400 V. Machines that use capacitors as a means to start them, after which the capacitor is automatically disconnected from the system, are usually of the 230 working volt type. Should the latter be connected into circuit with the former, a breakdown can be expected.

Releasing dielectric energy
It has been shown that a capacitor's energy or electrostatic charge is stored within the dielectric material sandwiched between the two plates. A simple way to demonstrate the release of this energy is to short circuit the two disconnected leads serving the capacitor. The expanded orbits of the negatively charged dielectric electrons then cease to be under any stress and the electrons stored in the capacitor's negative plate are attracted towards the positive plate at enormous speed. This results in the familiar electrostatic 'crack' and illuminated display when the leads are placed together.

A silly and dangerous practice, which seems to stem mainly from apprentices or young electricians, is to charge a capacitor by means of a high working voltage, then to hand it to an unsuspecting individual. This is extremely dangerous as capacitors are able to store a great deal of energy and it has been known for the accumulated stored power to cause the victim's heart to go into fibrillation. The golden rule therefore must be:

DON'T PLAY WITH ELECTRICITY — IT CAN KILL!

Capacitors and alternating current

Unlike direct current (DC), alternating current (AC) is constantly reversing in polarity many times a second. A cycle is represented by one complete change of electrical polarity, often referred to as the *periodic time*. Generating companies within the United Kingdom develop a voltage wave form of 50 cycles per second, whereas in the United States of America the frequency is maintained at 60. This is written as 60 Hz (60 hertz), and phonetically pronounced as *'hurts'*. Named after *Heinrich Hertz* (1857–94), periodic wave form can be generated in many different shapes. the simplest, developed by rotating a single loop of wire within a uniform magnetic field, produces the familiar sinusoidal wave form pattern (Figure 5.12, later in the chapter).

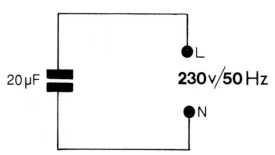

Figure 5.4. An alternating voltage is applied to a simple capacitor.

Applying an AC voltage to a capacitor as depicted in Figure 5.4 causes the plates alternately to charge and to discharge several times a second. When fully charged, the voltage stored is equal to the voltage administered.

Reactance

As the potential across the capacitor increases, it offers a resistance to current flow. Since the flow of current is entirely dependent on this 'counter-voltage', or 'capacitive voltage', and not resistance, the capacitive voltage is known as *reactance*. The electrical symbol for reactance is X_C and is known as *capacitive reactance*. Expression [4.10] shows that the capacitive reactance is the reciprocal of twice the value of π multiplied by the frequency of the supply in hertz and the value in farads of the capacitor under review. Hence

$$X_C = \frac{1}{2\pi fC} \qquad [4.10]$$

Unit of measurement
Capacitors are measured in *farads*, which are too large for practical purposes, so microfarads are used. A microfarad is one millionth of a farad and its electrical symbol is μF. Take as an example a small capacitor whose measured value is 0.000 02 farads. In order to convert to microfarads this must be multiplied by 1 million:

0.000 02 farads × 1 000 000
= 20 microfarads

To revert to farads, divide the value in microfarads by 1 million or move the decimal

point six places to the left. Smaller, too, is the picofarad, generally found in electronic circuits. It is equivalent to one-millionth of a microfarad, or 10^{-12} farad, and is expressed by the symbol pF.

Evaluating the current drawn
Figure 5.4 illustrates a simple capacitive circuit comprising a 20 μF capacitor connected in parallel formation to a 230 V, 50 Hz supply. To evaluate the current drawn from the circuit the capacitive reactance, X_C, must first be calculated.

Referring back to Expression [4.10],

$$X_C = \frac{1}{2\pi fC}$$

Substituting figures,

$$X_C = \frac{1}{2 \times 3.142 \times 50 \times 0.000\ 02}$$

$$X_C = \frac{1}{0.006\ 284}$$

$$X_C = 159.134 \text{ ohms}$$

Once the capacitive reactance has been calculated, the current flowing may be found by use of the following expression:

Current in amps

$$= \frac{\text{voltage}}{\text{capacitive reactance}} \qquad [5.1]$$

Alternatively, Expression [5.1] may be written as:

$$I = \frac{E}{X_C}$$

where E is the electromotive force in volts and X_C the capacitive reactance in ohms.

Substituting for known values,

$$I = \frac{230}{159.134}$$

$$\therefore I = 1.445 \text{ amps}$$

A less precise method A practical method to evaluate the working value of a capacitor in a pure capacitative circuit can be demonstrated by recording the values of both voltage and current and employing the following expression:

microfarads (μF)

$$= \frac{3182.6801 \times \text{current}}{\text{voltage}} \qquad [5.2]$$

where 3182.6801 is a constant factor.

A word of warning! This expression is only suitable for calculations involving supply systems generated in a 50 cycle per second wave form. For 60 Hz, the factor 2650 should be applied.

By way of illustration let us reconsider the example as shown as Figure 5.4:

$$\mu F = \frac{3182.6801 \times 1.445}{230}$$

$$\mu F = 19.99$$

A greater degree of accuracy may be achieved by calculating to the seventh decimal place. Variations of this expression can also establish the voltage and current in the circuit. Hence:

$$\text{current} = \frac{\text{microfarads} \times \text{voltage}}{K} \qquad [5.3]$$

or

$$\text{voltage} = \frac{K \times \text{current}}{\text{microfarads}} \qquad [5.4]$$

where K (the constant) is 3182.6801 (at 50 Hz)

The current flowing in a pure capacitive circuit will seem to travel from one plate to another. This is illusionary as both surfaces are completely isolated from each other by the dielectric insulation sandwiched between the plates. When an alternating current is applied to the capacitor, it is constantly charging and discharging every electrical cycle and it is this phenomenon which creates the illusion that the current is flowing through the capacitor.

Capacitors connected in series
Figure 5.5 illustrates how a 10 and 20 μF capacitor are connected in series formation. To calculate the total capacitance in the circuit, the following expression is used:

$$\frac{1}{C_t} = \frac{1}{C_1} + \frac{1}{C_2} \dots \qquad [5.5]$$

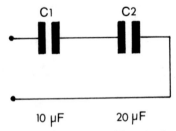

Figure 5.5. Capacitors connected in series formation.

where C_t is the total capacitance of the circuit in farads or microfarads and
  C_1 and C_2 are the individual values for each capacitor (common units must be used).

This expression is exactly the opposite to calculating resistances connected in series formation and therefore easy to remember.

As with evaluating resistances in parallel, the lowest common multiple (LCM) must first be found. Substituting figures,

$$\frac{1}{C_t} = \frac{1}{10} + \frac{1}{20} \quad (\text{LCM} = 20)$$

$$\frac{1}{C_t} = \frac{2 + 1}{20}$$

$$\frac{1}{C_t} = \frac{3}{20}$$

By cross-multiplying,

$$3C_t = 20$$

and dividing throughout by 3 to bring the expression in terms of C_t,

$$C_t = \frac{20}{3}$$

$$C_t = 6.66 \ \mu F$$

Another way to approach this problem is by applying the principle that:

$$C_t = \frac{C_1 \times C_2}{C_1 + C_2} \qquad [5.6]$$

Enumerating,

$$C_t = \frac{10 \times 20}{10 + 20} = \frac{200}{30}$$

$$C_t = 6.66 \ \mu F$$

This expression is only suitable for *two* capacitors connected in series formation. Should the total capacitance of more than two components be required, the traditional method shown as Expression [5.5] must be used.

Capacitors connected in parallel formation

Figure 5.6 illustrates how 10 and 20 μF capacitors are connected in parallel formation. The total capacitance of the circuit can be calculated by use of the following expression:

$$C_t = C_1 + C_2 \ldots \qquad [5.7]$$

where C_t represents the total capacitance of the circuit in farads or microfarads and C_1 and C_2 are the individual values for each capacitor (common units must be used).

An easy way to remember this expression is that it is the opposite of resistances connected in parallel formation. Substituting figures,

$$C_t = 10 + 20$$
$$C_t = 30 \ \mu F$$

Capacitors connected in both series and parallel formation

Figure 5.7 shows a capacitive circuit whose components are connected in both series and parallel formations. To calculate the total value of the circuit we must first consider the series leg of the circuit, summarised by Expression [5.6]. Once the total capacitance of the series configuration has been resolved it may then be reintroduced into the

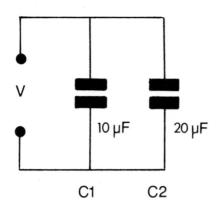

Figure 5.6. Capacitors wired in parallel formation.

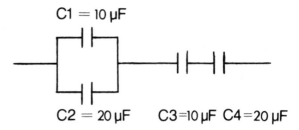

Figure 5.7. Capacitors wired in both series and parallel formation.

original circuit as a single item. By use of Expression [5.7] the remaining parallel section may be calculated to a single value. The total capacitance can then be established by applying Expression [5.6] to the two remaining capacitors connected in series formation.

Let us now evaluate the total capacitance mathematically: First, the total series leg of the circuit:

$$C_t = \frac{C_1 \times C_2}{C_1 + C_2} \qquad [5.6]$$

$$C_t = \frac{10 \times 20}{10 + 20}$$

$$C_t = 6.66 \ \mu F$$

Next, the total parallel section of the circuit:

$$C_t = C_1 + C_2 \qquad [5.7]$$
$$C_t = 30 \ \mu F$$

Finally the total capacitance of the two remaining components, connected in series formation, by use of Expression [5.6]:

$$C_t = \frac{6.66 \times 30}{6.66 + 30}$$

$$C_t = 5.450 \ \mu F$$

Practical applications

A faulty capacitor will obviously require a replacement which is of the correct working voltage and value. If the precise size is unavailable as a single component, the equivalent may be designed by employing other capacitors connected in either series or parallel configuration or a mixture of both.

As a practical example, consider a small electric motor served by a faulty 4 μF oil-filled capacitor. By connecting two capacitors in series formation the correct value may be obtained. This can be theoretically calculated by use of the following expression:

$$\frac{1}{C_t} = \frac{1}{C_1} + \frac{1}{C_x} \qquad [5.8]$$

where C_t is the total capacitance required,
$\quad\quad$ C_1 the value of an available capacitor and
$\quad\quad$ C_x is the unknown value of the accompanying capacitor.

Let the value of one of the available capacitors be 20 μF. Substituting for figures known,

$$\frac{1}{4} = \frac{1}{20} + \frac{1}{C_x}$$

Expressing in terms of C_x,

$$\frac{1}{C_x} = \frac{1}{4} - \frac{1}{20}$$

Finding a common multiple (20):

$$\frac{1}{C_x} = \frac{5 - 1}{20}$$

$$\frac{1}{C_x} = \frac{4}{20}$$

By cross-multiplying and bringing in terms of C_x,

$$C_x = 5 \ \mu\text{F}$$

By calculation, to produce a replacement value of 4 μF it is necessary to connect 5 and 20 μF capacitors in series formation. This would then bring the total capacitance to 4 μF.

Circuits containing both capacitance and resistance

An interesting point to mention at this juncture is that no power is consumed in a pure capacitive circuit. Both current and voltage are out of phase with each other by 90 degrees and the energy stored in the capacitor is returned to the supply at the end of each half cycle.

We have seen how the entire value of a group of resistors or capacitors can change according to

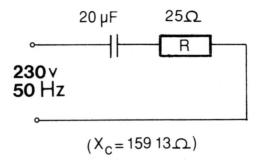

$$(X_C = 159 \ 13 \ \Omega)$$

Figure 5.8. A circuit with an element of both capacitance and resistance.

how they are connected. Should a resistance be added in series formation to a pure capacitive circuit, illustrated in Figure 5.8, a new electrical factor will emerge. The current flowing in such a circuit is limited to the capacitive reactance, X_C, of the capacitor and the resistance, R, in ohms of the resistor. These two items together form what is known as the *impedance* of the circuit and is measured in ohms. The symbol for impedance is Z, and can be calculated by applying the following expression:

$$Z = \sqrt{(R^2 + X_C^2)} \qquad [5.9]$$

where R $\;$ is the resistance of the circuit in ohms
$\quad\quad$ and
$\quad\quad$ X_C (or $1/(2\pi fC)$ Expression [4.10] is the capacitive reactance in ohms.

Substituting for figures given:

$$Z = \sqrt{25^2 + 159.13^2}$$
$$Z = \sqrt{625 + 25\ 322.35}$$
$$Z = 161.0 \ \Omega$$

We have seen how capacitors interact with resistors when connected in series formation and have mathematically calculated the total impedance offered to a voltage applied to the circuit. A daunting task to say the least, especially if the values are large and your pocket calculator batteries have just run out!

The third and final factor to be considered is the *inductor*, and we shall be dealing with this next.

Inductive circuits

A pure inductive circuit uses a coil of wire in much the same way as a resistor would be used to

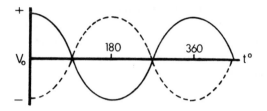

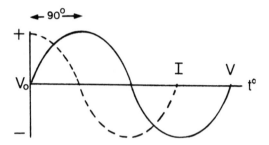

— Applied voltage
-- Induced emf

Figure 5.12. A secondary voltage induced into an inductor directly opposes the polarity of the voltage creating it.

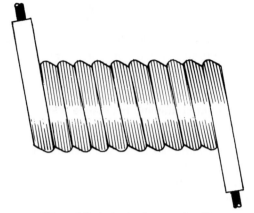

Figure 5.9. A simple air-core solenoid.

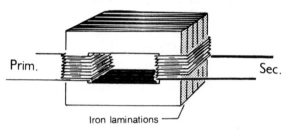

Figure 5.10. A simple transformer.

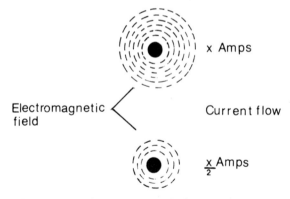

Figure 5.13. In an inductive circuit the current lags behind the voltage by 90 degrees, whereas in a pure capacitive circuit the current leads the voltage by 90 degrees.

Figure 5.11. The electromagnetic field surrounding a load-bearing conductor is directly proportional to the current flowing.

provide a simple load. Figures 5.9 and 5.10 illustrate two examples of an inductive circuit: the air-core solenoid and a small step down transformer. Practical applications include auto-transformers, electric bells, electromagnets and electric motors.

An electrical load-bearing conductor is accompanied by an electromagnetic field which surrounds the conductor (Figure 5.11). When this concept is applied to an inductor it has the effect of inducing a secondary voltage in the windings, directly opposing the polarity of the voltage creating it. This may be seen more clearly by studying Figure 5.12. Since the secondary, or induced, emf is 180 degrees out of phase with the applied voltage it will act as a current limiter. Although acting very similarly to a typical resistance, it is not a resistance but is caused through inductance and is known as *inductive reactance*. The unit of inductance is called the *henry* (H), identified by the symbol X_L, and is quantified in ohms.

The value of an inductive circuit may be calculated by use of the following expression:

$$X_L = 2\pi f L \qquad [5.10]$$

where f is the frequency of the voltage applied in hertz and
   L the value of the inductor in henries.

In a pure inductive circuit the current lags behind the voltage by 90 degrees as Figure 5.13 shows. The stored energy is within the electromagnetic field surrounding the inductor and is returned to the circuit each time the magnetic field collapses. Less energy is induced to the circuit, as a proportion is dissipated in the form of heat generated to overcome the natural resistance of the coil winding.

Once the value of the inductive reactance has been found, it may be used to calculate the total current flowing in the circuit by the expression:

$$I = \frac{E}{X_L} \qquad [5.11]$$

where I is the total current in amps,
E the value of the applied AC voltage and
X_L is the inductive reactance in ohms.

Alternatively, Expression [5.11] may be expressed as:

$$I = \frac{E}{2\pi fL} \qquad [5.12]$$

As a practical example, consider the following problem:

Calculate the value of the inductive reactance in ohms and the current consumed in amps when a 0.382 henry locking solenoid is connected to a 230 V, 50 Hz supply.

First the inductive reactance X_L must be calculated:

$$X_L = 2\pi fL \qquad [5.10]$$

Substituting figures,

$$X_L = 2 \times 3.142 \times 50 \times 0.382$$
$$X_L = 120.024 \ \Omega$$

Next, the current consumed must be calculated:

$$I = \frac{E}{X_L} \qquad [5.11]$$

Substituting for known values,

$$I = \frac{230}{120.024}$$

$$I = 1.916 \ \text{amps}$$

An interesting point to consider at this stage is had the frequency been 60 Hz, the inductive

reactance would have calculated to be 144.02 Ω and the current consumed only 1.59 A. The higher the frequency, the greater the inductive reactance and therefore the current flow would be proportionally lower.

Series circuit containing resistance, capacitance and inductance

Figure 5.14 illustrates a simple series formation circuit containing one 10 μF capacitor, a 0.38 H solenoid and a 10 Ω resistor connected to a 230 V, 50 Hz supply.

The total impedance, Z, of a series circuit comprising all three components can be calculated by use of the following expression:

$$Z = \sqrt{[R^2 + (X_L - X_c)^2]} \qquad [5.13]$$

where R is the total resistance in ohms,
X_L is the value of the inductive reactance in ohms and
X_C is the capacitive reactance in ohms.

In order to calculate the total impedance, the value of the inductive and capacitive reactance must first be found. Firstly, the inductive reactance:

$$X_L = 2\pi fL \qquad [5.10]$$
$$X_L = 2 \times 3.142 \times 50 \times 0.382$$
$$X_L = 120.024 \ \Omega$$

Next, the capacitive reactance:

$$X_C = \frac{1}{2\pi fC} \qquad [4.10]$$

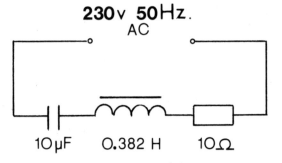

Figure 5.14. A series formation circuit comprising a capacitor, solenoid and resistor.

$$X_C = \frac{1}{2 \times 3.142 \times 50 \times 0.000\ 01}$$

$$X_C = 318.268\ \Omega$$

The values of all three components have now been evaluated:

R = 10 ohms (taken directly from Figure 5.14)

X_L = 120.024 ohms (inductive reactance)

X_C = 318.2 ohms (capacitive reactance)

Returning to Expression [5.13] and substituting figures:

$$Z = \sqrt{[10^2 + (120 - 318.2)^2]}$$
$$Z = \sqrt{[100 + (-198.2)^2]}$$

(Remember that a minus quantity multiplied by another minus quantity produces a positive resolution.)

$$Z = \sqrt{39\ 383.24}$$

Therefore

$$Z = 198.4\ \Omega\ \text{impedance}$$

The current consumed may now be calculated by use of the following expression:

$$I = \frac{E}{Z} \qquad [5.14]$$

where E is the applied emf in volts.

$$I = \frac{230}{198.4}$$

$$I = 1.15\ \text{amps}$$

The collective volt drop

A curious aspect of this type of problem is that the collective volt drop across all three components cannot be added arithmetically as, for example, when measurements are calculated with DC circuits. Answers and verification to AC series formation circuits must not be proved in this manner.

To simplify:

Voltage across the resistor (IR)
= 1.15 × 10 = 11.5 V

Voltage across the inductor (IX_L)
= 1.15 × 120 = 138 V

Voltage across the capacitor (IX_C)
= 1.15 × 318.2 = 365.93 V

Power generated with the circuit To calculate the total power generated, Expression [2.9] must be called upon:

Power in watts = I^2R

Since no power is expended in either the capacitor or inductor, as energy levels are returned to the circuit every cycle,

$$W = 1.15^2 \times 10$$

Therefore,

power in watts = 13.2

Expression [5.13] is not only useful to calculate the total impedance of a series circuit containing resistance, capacitance and inductance, but also forms a corner-stone for other electrical AC computations. As an aid to finding solutions to theoretical problems, this expression is extremely useful and it is worthwhile spending both time and effort to understand fully the concepts involved.

Capacitor types

There are many varieties of capacitors to be found in general use today. Some have specialised responsibilities and are used for electronic assignments. Others are far more versatile and are applied to more general electrical engineering. Six of the more popular types of capacitors which are commonly used today are now described.

Air capacitor

This type of capacitor has movable plates, using the air as a dielectric. These capacitors may be found in radio circuits for tuning purposes. Values range from 10 to 1000 pF.

Paper capacitor

Constructed by means of two long strips of thin aluminium foil, this variety of capacitor is served by a dielectric made from waxed paper. Once positioned between the two plates, the complete assemblage is rolled and placed within a suitable container. A termination point, for wiring purposes, is anchored to each plate. Paper-insulated capacitors vary in value from a few picofarads to many microfarads and are used in low frequency electrical installation work. Examples may be drawn from motor circuits employing capacitors as a means of starting and supply problems relating to improved power factors.

Ceramic capacitor

Ceramic capacitors are generally used for high frequency work such as radio, television and radar. This type of capacitor normally takes the form of a ceramic disc or tube sandwiched between two silver foil plates.

Electrolytic capacitor

This variety has a very high capacitance for its size but can break down easily — a point well worth remembering! Two strips of aluminium foil, with gauze containing ammonium borate, are rolled to form a cylindrical component and housed in a suitable enclosure. An oxide insulating film is deposited on one of the plates and acts as a dielectric. The electrolytic capacitor is polarised, having both positive and negative terminals. Extreme care should be taken to ensure that the capacitor is correctly installed, otherwise damage could result. The electrolytic capacitor will break down should the voltage supplied exceed its dielectric value. It is helpful to recognise at this stage that the dielectric strength is not proportional to thickness; so doubling the thickness of the dielectric will not provide additional protection against breakdown.

This type of capacitor is often found in DC smoothing circuits and may also be used in DC power supplies. A small but continuous leakage is experienced and may only be applied to circuits where the polarity is unchangeable. Values and polarity are usually specified, together with the working voltage and maximum temperature rating.

Mica capacitor

Mica, acting as a dielectric, has a thin film of silver deposited directly to its surface. Plates are then arranged in parallel formation and packaged in a suitable container. It has low loss and high dielectric strength so can be considered to be very stable when compared with the electrolytic variety. It is used mainly in television and high frequency circuits, and its capacitance can vary from a few picofarads to approximately 0.01 μF.

Oil-impregnated paper capacitor

An oil-impregnated paper, used as a dielectric material, is sandwiched between two strips of aluminium foil. Similarly constructed to the waxed paper capacitor, it usually takes the form of a rolled or oval assembly.

Capacities fluctuate between 40 pF to many microfarads. It is mainly used for low frequency work.

Summary

We have seen how the simplest form of capacitor can be constructed by use of two metal plates separated by a special form of insulating material called a dielectric. Different materials were shown to have contrasting dielectric strengths, of which several were listed.

Next, theoretical concepts were studied and it was demonstrated why the orbits of the electrons, within the dielectric material, were stretched and attracted to the positive plate when fully charged.

Problems involving capacitors connected in series and parallel formation were considered. Theoretical methods to determine total capacitance were found to be, in every respect, opposite to procedural guidelines governing resistances connected in a similar manner.

Alternating voltage applied to a capacitor was shown to offer resistance to current flow. This was known as capacitive reactance. The circuit was then expanded to include an element of resistance, after which inductance was introduced. Calculations were then made to evaluate the total impedance of the circuit, identified by the symbol Z:

$$Z = \sqrt{[R^2 + (X_L - X_C)^2]} \qquad [5.13]$$

To recapitulate:

capacitive reactance $(X_C) = \dfrac{1}{2\pi fC}$ [4.10]

inductive reactance $(X_L) = 2\pi fL$ [5.10]

We have examined how capacitors can hold their charge and then return to a neutral state. Installation cables, under certain conditions, can also hold a capacitive charge, and it is this we will be deliberating within the next chapter when consideration will be given to cables and theoretical wiring methods.

6 Cables and theoretical circuitry

In this chapter: A brief history. Volt drop and temperature variations. Factors affecting the resistance of cable. Calculations and wiring systems. Power and final circuit arrangements. Theoretical methods of testing. Fibre optic cables.

Former times

The first land telegraph cable was laid in 1844 when houses were lit by gas and oil lamps. At that time electricity as a source of power was no more than a scientific curiosity. It was to be 37 years later, in 1881, that the first public cable was commissioned. The cable consisted of solid drawn 'D' shaped copper bars supported on insulators and drawn into a steel conduit which was filled with bitumen. A crude attempt compared with today's standards, but it was successful and proved to be the corner-stone of cable design to come.

Many early electrical installations were carried out using a technique known as *cabinet casing*. In those days being an electrician involved basic woodworking skills, as lighting and power cables were laid within lengths of grooved timber. Cabinet casing came in various sizes from 30 to 60 mm in diameter, ($1\frac{3}{16}$ to $2\frac{3}{8}$ of an inch), but the overall depth remained constant at approximately 10 mm (see Figure 6.1).

Cable joints were made by physically joining the cables together and soldering. A layer of black fabric-based insulating tape was then applied to the joint. To seal the installation a thin timber lid was tacked on to the grooved timber base. This technique can still be seen, although redundant, in very old properties where rewiring has taken place and the old installation left.

Cables wrapped in impregnated tape and sheathed in lead or flat rubber were widely used before the advent of the Second World War and remained in popular use up to the early 1950s.

Polyvinyl chloride (PVC) cables and minerally insulated (MI) cables were very slow to gain acceptance, but are now two of the most widely used cables in electrical installation work.

Cable resistance and the effects of volt drop and temperature variations in a circuit

All cables have a natural intrinsic resistance and under certain defined conditions can be the cause of considerable damage to both installation and, in some cases, the appliance used.

The natural resistance of any cable may be calculated by using the expression:

$$R = \frac{\rho l}{a} \qquad [6.1]$$

where ρ is the resistivity of the conducting material in ohms per metre (Ω/m) and is pronounced 'rho' — the resistivity for copper, at 20 °C, has been calculated to be 1.8×10^{-8} Ω/m for aluminium, 2.8×10^{-8} Ω/m,

l is the length of the cable in metres,

a is the cross sectional area in metres squared (m^2) and

R is the total resistance of the cable in ohms.

Natural cable resistance will cause a volt drop to occur within the conductor and is directly proportional to the current flowing in the circuit. This may be calculated by use of Expression [2.7]:

volt drop $= I \times R$

where I is the current flowing in the circuit and

R is the natural resistance of the cable in ohms.

As an example, consider Figure 6.2 and the following:

Figure 6.1. Cabinet casing.

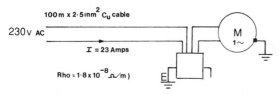

Figure 6.2. A 230 V induction motor drawing 23 A from the supply is served with 100 m of 2.5 mm² copper cable.

A 200 m circuit of copper cable supporting a cross-sectional area of 2.5 mm is installed to serve a single phase, 230 V induction motor drawing 23 A from the supply. Assume an ambient temperature of 20 °C.

By applying Expressions [2.7] and [6.1] in tandem, the total volt drop may be calculated. The resistivity of copper at 20 °C is 1.8×10^{-8} Ω/m.

$$\text{volt drop} = I \left(\frac{\rho l}{a}\right) \qquad [6.2]$$

Substituting figures and remembering to bring all

terms to metres in order that the expression may be calculated in standard units:

$$1 \text{ m}^2 = 1000 \times 1000 \text{ mm}$$
$$= 1\,000\,000 \text{ mm}^2$$

or

$$1 \text{ mm}^2 = 1 \times 10^{-6} \text{ m}^2$$

$$\text{volt drop} = 23 \left(\frac{1.8 \times 10^{-8} \times 200}{2.5 \times 10^{-6}}\right) \quad [6.2]$$

$$\text{volt drop} = 23 \times 1.44$$
$$\therefore \text{ VD} = 33.12 \text{ volts}$$

Current flow and resistance

Had the load been replaced with a 500 W quartz halogen lamp, drawing 2.173 A from the supply, the volt drop would have been reduced to 3.12 V. This would represent a drop of 1.35 per cent when connected to a 230 V supply and clearly shows the relationship between the current flowing and the intrinsic resistance of the cable. Figure 6.3

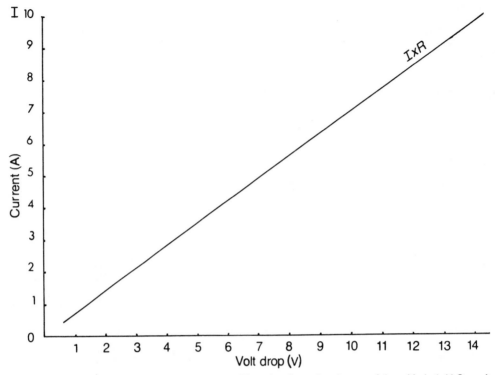

Figure 6.3. The relationship between volt drop and current. Given that the total resistance of the cable is 1.44 Ω a volt drop of 5.76 V would be experienced when a current of 4 A is flowing in the circuit.

illustrates the correlation between a variable current, volt drop and the natural resistance of a cable.

In practical terms, an electric motor experiencing such a large volt drop would activate the thermal overload circuit. If protection was set too high, or the machine fitted without an overcurrent device, the motor would most probably burn out.

It is important not only to select the correct sized cable to suit the load, but also to take into consideration the *length* of the circuit from the mains distribution centre to the appliance.

Practical volt drop
Another simple but practical method in which a volt drop can be demonstrated is by employing two 1.5 V batteries, connected in series formation to serve an electric bell. Should the installation be carried out using standard bell wire and a control button placed some 30 or 40 m away, the volt drop experienced within the conductors will be of such a magnitude as to silence the bell.

Under these conditions it would be better to incorporate a suitable 6−8 V transformer in circuit as shown in Figure 6.4 (see Chapter 7).

Factors affecting the resistance of a conductor

There are four factors affecting the resistance of a conductor and these may be summarised as follows:

1. The composition of the conductor.
2. The cross-sectional area of the conductor.
3. The temperature of the conductor.
4. The length of the conductor.

Material composition
In Chapter 1 it was demonstrated that electrical conductivity is very dependent on the size of the atom and the number of valance electrons in its outer shell. Silver is one of the best conductors of electricity followed closely by copper. These materials have large atoms and just one loosely

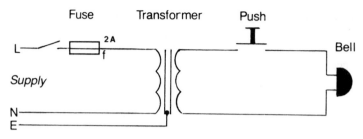

Figure 6.4. A simple bell circuit incorporating a step down transformer.

held valance electron. For practical purposes only copper and aluminium are used in the manufacture of power cables, silver being far too expensive.

Resistivity Different materials, comparable in both size and length, show contrasting values of electrical resistance. To bear comparison, the term *resistivity* is used and is *the resistance between two opposite faces of a metre cube of the material under examination.* The unit of resistivity is the ohm/metre, identified by the Greek letter ρ.

Cross-sectional area

The resistance of a conductor is inversely proportional to its cross-sectional area (csa). As an example take two water pipes, one large and one small, as Figure 6.5 illustrates. Let the water pressure in the tank represent the voltage, and the flow of water through the pipes, the current in the circuit. The natural resistance of the circuit may be compared with the diameter of the water pipes.

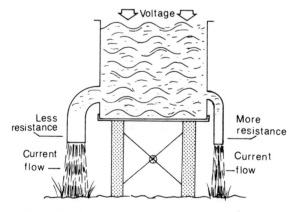

Figure 6.5. Far less resistance is experienced in the larger water pipe compared with its smaller companion.

Figure 6.5 clearly demonstrates that the larger of the two pipes offers far less resistance than the smaller one. The larger pipe enables a greater volume of water to flow, and so it is with electricity. The larger the cross-sectional area of the cable, the smaller the resistance will be when comparing with another of a lesser size.

In practical terms, if the current is excessively high and the conductor too small for the connected load, the cable will warm up. It is important to select the correct sized cable at the design stage of the installation, taking into consideration the proposed connected load and the effect volt drop might have on the circuit.

Temperature

The third factor affecting the resistance of a conductor is temperature and this must be taken into account when evaluating the expected change in cable resistance due to extreme temperature differentials. Generally, a rise in temperature will produce a *rise* in resistance for most conductors. However, there are one or two elements that have a *negative coefficient of resistance* and will respond with a *decrease* in resistance to an increase in temperature. Carbon and silicon are two examples of basic elements which display these characteristics.

To evaluate the changed resistance of a cable subjected to a rise in temperature, the following expression may be used:

$$\frac{R_1}{R_2} = \frac{1 + (\alpha t_1)}{1 + (\alpha t_2)} \qquad [6.3]$$

where R_1 is the initial resistance of the cable,
$\quad\quad R_2$ the final resistance of the cable,
$\quad\quad t_1$ is the initial temperature, in degrees Celsius (°C),

t_2 is the final temperature, in degrees Celsius and

α (the Greek letter alpha), is the temperature coefficient of the conductor.

For copper, α is calculated at 0.0043 ohms for each degree Celsius rise whereas tungsten has a temperature coefficient of 0.0045 $\Omega/°C$. This factor is often written as $45 \times 10^{-4} \ \Omega/°C$.

Converting to degrees Celsius Should the temperature be expressed in degrees Fahrenheit the following expression may be used to convert to Celsius:

$$°C = \frac{(°F - 32) \times 5}{9} \qquad [6.4]$$

If expressed in degrees Kelvin:

$$°C = K - 273.15 \qquad [6.5]$$

As a practical example, consider Figure 6.6.

A 230 V single phase circuit, wired in 2.5 mm^2 minerally insulated copper cable is found to be drawing 4.31 A from the supply. The installation is environmentally placed where the ambient temperature varies from 5 to 30 °C over a 24 hour period. At 5 °C the resistance of the cable is found to be 1.276 Ω. Evaluate both resistance and volt drop within the cable at 30 °C.

The temperature coefficient of copper is taken to

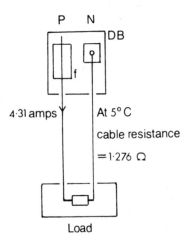

P   N

DB

f

4·31 amps

At 5° C

cable resistance

= 1·276 Ω

Load

Figure 6.6. A single phase circuit where the ambient temperature varies from 5 to 30 °C.

be 0.0043 ohms per degree Celsius. To calculate the resistance of the cable at 30 °C, reference must be made to Expression [6.3]:

$$\frac{R_1}{R_2} = \frac{1 + (\alpha t_1)}{1 + (\alpha t_2)}$$

Substituting figures,

$$\frac{1.276}{R_2} = \frac{1 + (0.0043 \times 5)}{1 + (0.0043 \times 30)}$$

Cross-multiplying,

$$1.0215 \times R_2 = 1.276 \times 1.129$$

Dividing throughout by 1.0215:

$$R_2 = \frac{1.276 \times 1.129}{1.0215}$$

$$R_2 = 1.41 \text{ ohms}$$

Increase in resistance due to rise in temperature

$$= 1.41 - 1.276 \text{ ohms}$$
$$= 0.134 \text{ ohms}$$

Evaluating voltage drop due to temperature variations Regulation 525-01-02 clearly stipulates that voltage drop between the incoming supply terminals and the fixed current-using equipment should not exceed 4 per cent of the nominal voltage (U_0) of the supply.

This may be calculated by use of the following expression:

$$\text{Permissible volt drop} = U_0 \times 0.04 \qquad [6.6]$$

where U_0 is the nominal voltage and 0.04 the percentage voltage drop permitted.

A maximum drop, disregarding electric motor starting periods, of 9.2 V is permissible within a 230 V supply system but should a 110 V arrangement be adopted the acceptable voltage drop would be reduced to 4.4 V.

Referring back to the practical example, Figure 6.6 and summarising:

The current flowing in the circuit was found to be 4.31 A. The natural resistance of the cable at 5 °C was given to be 1.267 Ω.

By applying Expression [2.7] the volt drop at 5 °C may be calculated:

$$\text{volt drop} = I \times R_1$$

Substituting figures

$$\text{volt drop} = 4.31 \times 1.276$$
$$= 5.49 \text{ volts}$$

Expression [6.3] has shown that a temperature differential of 25 °C produced an increase of 0.134 Ω within the conductor; giving rise to an additional volt drop, IR, of 0.577 V to develop. R_2, the resistance of the cable at 30 °C, was calculated to be 1.41 Ω. By applying Expression [2.7] the volt drop at 30 °C may be found:

$$\text{volt drop} = IR_2$$

Substituting figures,

$$\text{Second volt drop} = 4.31 \times 1.41$$
$$= 6.077 \text{ volts}$$

The additional voltage reduction developed within the circuit due to an increase in cable temperature is theoretically not sufficiently high enough to breach the demands of Regulation 525-01-02.

In practical terms a cable such as this would be changed as the volt drop experienced due to the rise in temperature would be considered unacceptable and therefore marginal to the electrical regulations. This is an academic example, but provides an important insight into problems involving conductors which are subjected to temperature variations.

To summarise: a higher temperature differential will develop an increase in conductor resistance in the majority of cables. This in turn will produce an additional voltage drop within the circuit.

It is important to understand this electrical concept fully, as volt drop can be the cause of considerable anxiety to an operative who is confronted with this problem and is not sure where the solution may lie.

Length

The last factor affecting the resistance of a cable is length. Since ρ, the resistivity of a conducting material, is a constant throughout the entire

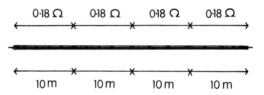

Figure 6.7. The resistance offered to a current is directly proportional to the length of the conductor.

conductor, it stands to reason that the resistance is in direct proportion to length, as Figure 6.7 shows.

By use of Expression [6.1] the resistance of a conductor may be found when the resistivity of the material is known. Electrical resistivity may be expressed as the total resistance in ohms across two opposite sides of a metre cube of the material under test. For copper, this figure has shown to be 1.8×10^{-8} Ω/m.

Practical methods The resistance of a conductor may be measured reasonably accurately by forming a loop from the cable under test and connecting an ammeter in series with one of the conductors. The two free ends are then connected to a 6 V battery and then made parallel with a voltmeter (Figure 6.8). The total resistance of the cable can then be calculated by use of Expression [1.4]:

$$R = \frac{V}{I}$$

As a practical example, consider the following:

A 50 m length of twin PVC insulated and sheathed lighting cable is joined together at one end in order to form a loop for the purpose of measuring the total resistance of the cable. An ammeter, connected in series with the circuit, records a current flow of 3.333 A. The voltage is supplied from a 6 V battery.

Referring back to Expression [1.4]:

$$R = \frac{V}{I}$$

and substituting for figures,

$$R = \frac{6}{3.333}$$

$$R = 1.8 \text{ ohms}$$

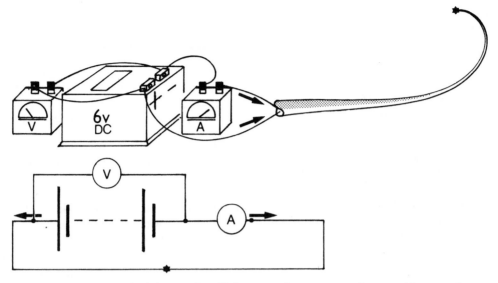

Figure 6.8. Measuring the resistance of a cable by means of an ammeter, voltmeter and battery pack.

Obviously there are far more accurate methods employed to measure the resistance of the cable. If the resistance of the cable under test is not known or a suitable ohmmeter is unavailable, this is a reasonable and practical method to adopt.

Theoretical wiring systems

Lighting circuit arrangements

Lighting circuits are usually wired from the nearest suitable distribution centre,using a cable appropriately designed to carry the envisaged load. In the UK, such an installation would be carried out using either 1.0 or 1.5 mm² cable. Domestic installations are often wired as economically as possible using PVC insulated and sheathed cables, whereas industrial systems can be wired using single insulated PVC cables drawn through solid drawn or welded steel conduit or trunking.

Figure 6.9 illustrates the basic circuitry required when installing a one-way lighting circuit. The symbols used are to British Standards, Number 3939 (BS 3939), a selection of which may be found in Appendix B.

Two-way switching Confusion can often arise when two-way switching is carried out. By its very nature the system is wired in an entirely different fashion from the familiar one-way arrangement.

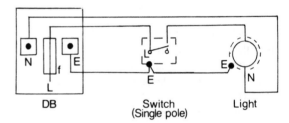

Figure 6.9. One-way lighting circuit.

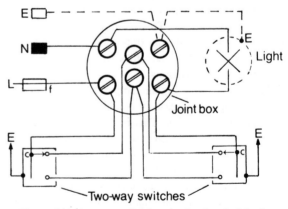

Figure 6.10. A two-way lighting circuit using the joint box method.

Figure 6.10 shows, schematically, how this might be carried out using the joint box method. In practice, the switching arrangement is wired using two and three core PVC insulated and sheathed

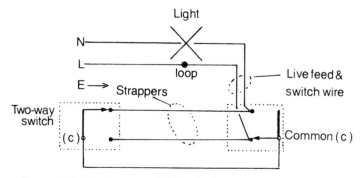

Figure 6.11. A basic one-way lighting circuit converted to a two-way circuit.

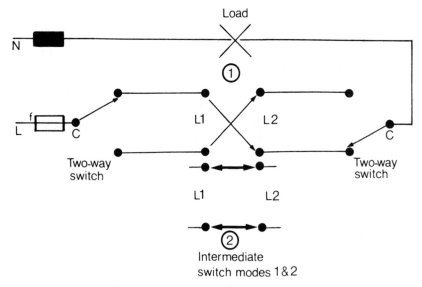

Figure 6.12. A basic two-way and intermediate switching arrangement.

cables for domestic installations, whereas single PVC insulated copper cables drawn through steel or plastic conduit are more suitable for industrial and commercial purposes.

An alternative technique used to wire a two-way lighting system without the use of a joint box can be confirmed by studying Figure 6.11. Sometimes known as the '*conversion method*', it has the advantage of having all cable connections accessible from one level and is often much quicker to install.

Intermediate switching Figure 6.12 illustrates the method used to incorporate intermediate switching into a two-way lighting circuit. As the name suggests, this type of arrangement employs a pair

of two-way switches, but there are no numerical constraints when intermediate switching is introduced into the circuit.

Both methods and techniques covering the practical side of lighting circuit arrangements will be dealt with in Chapter 9.

Power circuit arrangements
Socket outlets, designed to serve either domestic or commercial installations, are usually wired in the form of a *final ring circuit*. Once called a ring main, this type of installation has many advantages over the more conventional methods of wiring. Unlike the single socket radial circuit method, Figure 6.13, a final ring circuit can serve an unlimited number of sockets, provided certain

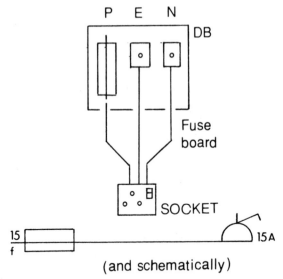

(and schematically)

Figure 6.13. A 15 A radial circuit.

criteria are met. These criteria concern the size and type of cable being used, together with the area which the installation has to serve. Ground rules are laid providing standards governing circuit overcurrent protection and the form which outlets may take in order to serve the final ring circuit. Figure 6.14 illustrates schematically the form such an installation might take (Regulation 314, Chapters 43, 46 and 55 refer).

Spurred sockets Spurs originating from socket outlets forming part of the final ring circuit provide a useful way of wiring to remote areas where it might be impractical to extend the cables forming the ring. In the UK, only one spurred socket is permitted per socket forming the final ring circuit. If 10 13 A socket outlets have been designed to form the main installation, an additional 10 may be added to the circuit in the form of spurred sockets. One spurred socket may be fed from each of the original main line sockets. Figure 6.15 will help make this more understandable.

As an exception to this rule, an unlimited number of socket outlets may be installed as a spur providing the spur is supplied by means of a fused connection unit. Figure 6.16 illustrates this point and shows that a maximum of 13 A may only be drawn from this section of the final ring circuit, no matter how many sockets are installed!

Permanently connected appliances may be connected to a final ring circuit providing they are served by a switched fused connection unit as shown in Figure 6.14.

It is important to bear in mind that the size of the overcurrent protection device attending the final ring circuit will be either rated at 30 or 32 A, so it would be impractical to incorporate

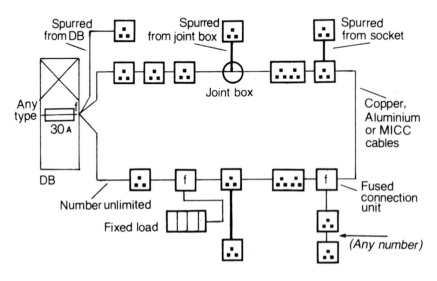

Figure 6.14. Final ring circuit. Recommended for a maximum floor area of 100 m². PVC, rubber insulated or MICC cables may be used. Conductor csa: copper, 2.5 mm²: aluminium, 4.0 mm²: MICC, 1.5 mm².

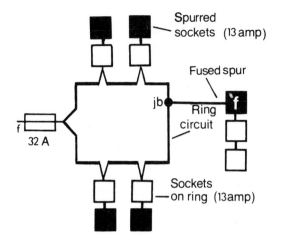

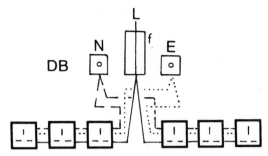

Figure 6.17. Two radial circuits wired in parallel at the point of origin can, at first sight, appear as though a final ring circuit has been wired.

Figure 6.15. One unfused spur may be wired from each main line socket outlet.

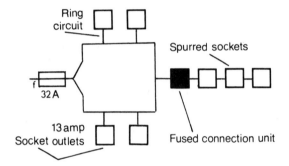

Figure 6.16. An unlimited number of 13 A socket outlets may be served by a fused connection unit.

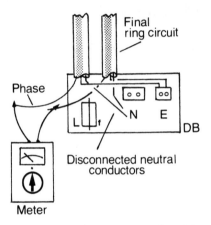

Figure 6.18. Testing the continuity of a final ring circuit: *step 1* — measuring the resistance of both phase and neutral conductors.

permanently connected appliances which are consumers of high current.

Testing Once a final ring circuit has been installed it must be tested for continuity and to establish that the circuit is a complete ring and not two radial circuits wired in parallel at the point of origin. Figure 6.17 will help to make this point more clear.

To test, both the phase and neutral conductors, together with the current protective cables (cpc), are disconnected from the distribution board, leaving six free cable ends. By using a continuity tester the resistance must first be established between the open ends of the conductors and noted, as illustrated in Figures 6.18 and 6.19.

At an approximate mid-point position in the circuit all three conductors are made common with

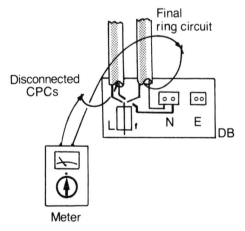

Figure 6.19. Testing the continuity of a final ring circuit: *step 2* — measuring the resistance of the current protective conductor.

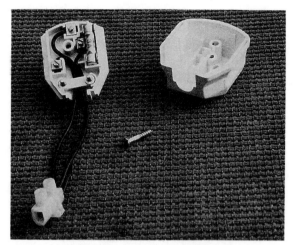

Figure 6.20. Linking out phase, neutral and current protective conductors by means of short circuiting a 13 A plug.

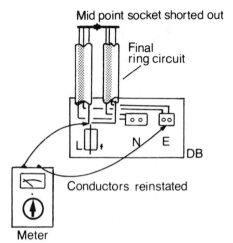

Figure 6.22. A final reading must be taken between the phase and current protective conductor.

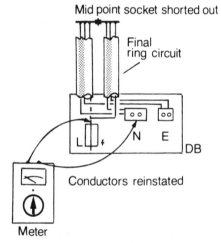

Figure 6.21. The value recorded should be approximately half of the recorded value at the start of the test.

each other. This can be accomplished by linking together the cables serving a socket outlet or joint box as shown in Figure 6.20. After this stage has been reached, the six free cable ends forming the beginning and end of the final ring circuit can be reinstated into the distribution board. A reading is then taken between the phase and neutral conductors. This should be approximately *half* the evaluation recorded at the commencement of the test when all the cable ends were open (Figure 6.21).

To conclude the test, a final reading must be taken between the phase conductor as Figure 6.22

illustrates. The value in ohms obtained should be a *quarter* of the original reading between the open ends of the phase conductor, *plus* a *quarter* of the reading taken from the open ends of the cpc.

The final ring circuit is then reinstated by removing the links which form the short circuiting facility.

In practice it is very difficult to select a mid-point position in the circuit in order to carry out a test, so all figures obtained tend to lead to an approximate resolution.

Practical application As a practical example, consider a final ring circuit serving five twin 13 A socket outlets over an area of 90 m^2.

First step A reading is taken between the open ends of the phase and neutral conductors and found to be 0.4 Ω.

Second step The test is then switched to the open ends of the current protective conductor and a reading of 0.8 Ω is recorded.

Third step After short circuiting the circuit at a mid-point position and re-establishing the open ends in the distribution board, a continuity test is taken between the phase and neutral conductors.

expected reading

$$\simeq \frac{1}{2} \times \frac{\text{1st step reading}}{1} \qquad [6.7]$$

Substituting figures,

$$R \simeq \frac{1}{2} \times \frac{0.4}{1}$$

$$R \simeq 0.2 \text{ ohms}$$

Fourth step To conclude the test, a final reading is taken between the phase conductor and the current protective conductor.

$$\text{expected reading} \simeq \frac{\text{1st step reading}}{4}$$
$$+ \frac{\text{2nd step reading}}{4} \qquad [6.8]$$

Substituting figures,

$$R \simeq \frac{0.4}{4} + \frac{0.8}{4}$$

$$R \simeq 0.1 + 0.2$$
$$R \simeq 0.3 \text{ ohms}$$

If figures obtained during testing meet the criteria laid down, then continuity of the final ring circuit has been proven. If unacceptably high, it might be possible that a loose cable termination is to blame by creating an exceptionally high impedance in the circuit. This should be dealt with immediately and retested after the fault condition has been found.

Post testing After completion, an insulation test must be carried out between all conductors. This is achieved by measuring the resistance, in millions of ohms, between the phase and neutral conductors and between a combination of both phase and neutral conductors to the principal current protective conductor. In the UK, the minimum acceptable value for a fixed wiring installation is 500 000 ohms, (0.5 megohm and written as 0.5 MΩ) If less, then the circuit should be checked for possible reasons to find why an anomaly has occurred (Regulation 713-04-04).

Figure 6.23 illustrates a step by step approach explaining how an insulation test may be carried out using a standard insulation test meter. The practical aspects of testing will be reviewed in Chapter 14. Remember, high voltage insulation testing can cause capacitance within the conductor. Once testing has been completed, discharge the

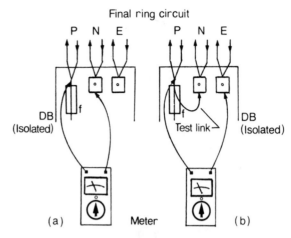

Figure 6.23. Insulation testing a final ring circuit.

cable by short circuiting to avoid unnecessary shock.

Radial circuits Where the maximum demand for current is small, the *radial circuit* is an acceptable alternative to the final ring circuit. Advantages stem mainly from cheaper installation costs.

Figure 6.24 shows, in schematic detail, the basic requirements for a radial circuit serving a floor area of approximately 50 m^2. As with the final

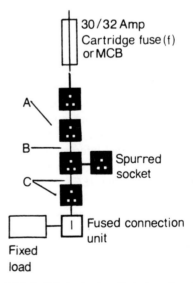

Figure 6.24. Radial circuit. A, unlimited number of 13 A socket outlets. B, cable size: PVC or rubber insulated — 4 mm^2; aluminium conductors — 6 mm^2; mineral insulated — 2.5 mm^2. C, maximum floor area of 50 m^2.

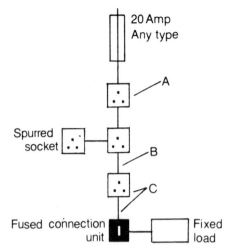

20 Amp
Any type

A

Spurred
socket

B

C

Fused connection
unit

Fixed
load

Figure 6.25. Radial circuit. A, unlimited number of 13 A socket outlets. B, cable size: PVC or rubber insulated — 2.5 mm²; aluminium conductors — 4 mm²; mineral insulated — 2.5 mm². C, maximum floor area of 20 m².

ring circuit, permanently connected equipment and an unlimited number of 13 A socket outlets may be added providing certain criteria are met (Regulation Chapters 43, 46 and 52). There are two types of radial circuits which may be wired. The first is designed to serve a floor area of up to 50 m², protected by an overcurrent device rated at 32 A. If wired using copper insulated cables, a minimum size of 4.0 mm² must be observed or 6.0 mm² when aluminium cables are chosen.

The second type of radial circuit provides for floor areas up to 20 m². The cable size is reduced to 2.5 mm² when employing copper conductors and 4.0 mm² when aluminium is used. Figure 6.25 summarises the criteria laid down for this type of radial circuit. Additional information may be obtained by referring to the current edition of the *Wiring Regulations* or the *IEE's On Site Guide*.

Final circuit arrangements: domestic
Wiring a domestic installation is usually a straightforward undertaking providing it is carried out whilst the property is being constructed, or when it is empty.

It is very easy to rush into a new job highly motivated with an aim of accomplishing as much productivity as possible in the minimum of time.

Often there are pressures outside our sphere of authority which trigger such an approach. An overbearing employer, pricing errors or inexperience are just three factors which might be responsible.

Blundering into a new job as if there were no tomorrow is not good strategy. Sit down and think the job through before starting work. Remember the acronym '*STRAW*' — *Stop, Think, Review and Work*.

Planning an installation There are many different circuits that may be included when planning an installation and some have been dealt with already. Listed are eight of the more commonplace circuits that have not been debated on previous pages. These circuits may be included in the design stage of a domestic installation:

1. Immersion heater circuit.
2. Central heating control circuit.
3. Cooker circuit.
4. Fixed wall heater circuit.
5. Electric shower circuit.
6. Door bell circuit.
7. Garage circuit.
8. Off-peak installation.

Consideration must be given to correction factors when determining the size of cable required, examples of which are reviewed in Appendix C.

The majority of the listed subjects can be classified as radial circuits. Each consists of one cable originating from the distribution board and is individually served with independent overcurrent protection means.

Immersion heater circuit Reference is made to Regulation 554-05. Figure 6.26 illustrates a typical circuit serving a 3 kW, 230 V domestic immersion heater. The current demand must be assessed before any installation is undertaken. In the UK the majority of domestic immersion heaters are rated at 3000 W and are designed to operate from a 230 V supply. By applying Expression [2.5], the total current may be calculated for this type of installation:

$$I = \frac{W}{V}$$

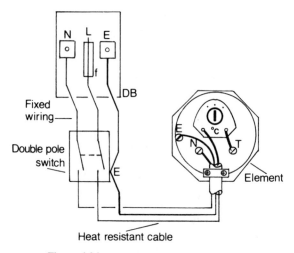

Figure 6.26. Typical immersion heater circuit.

Substituting figures,

$$I = \frac{3000}{230} = 13 \text{ amps}$$

Had the installation been carried out where the supply voltage was 110 V, the current demand would be:

$$I = \frac{3000}{110} = 27.272 \text{ amps}$$

Once the current has been assessed, a suitably sized cable can be selected from tables within the current edition of the Institute of Electrical Engineers (IEE) Wiring Regulations or in the *National Electrical Code* if work is undertaken in the USA. In the UK it is usual to wire an immersion heater circuit using 2.5 mm² PVC insulated and sheathed cable for domestic use. This cable is adequately sized for heating appliances of up to 4 kW.

Central heating control circuit Figure 6.27 outlines the basic wiring arrangements, in schematic form, for a central heating control supply. This may either be spurred from a final ring circuit, terminating in a fused connection unit, or served directly from a local distribution centre. Should the circuit originate from the source of the supply and be controlled by an independent overcurrent protection device, it must be terminated with a 20 A double pole switch to provide a means of isolation.

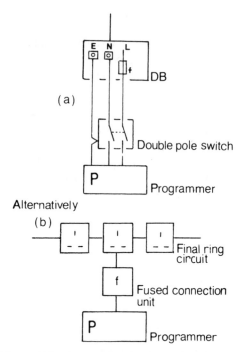

Figure 6.27. Alternative basic wiring arrangements for a central heating control supply.

The practical aspects of wiring control circuits and appliances serving a central heating installation will be dealt with in Chapter 12. As the current demand is so surprisingly low the system may be carried out using a lighting sized cable. Should there be any doubt, the total potential current drawn from the supply should be assessed and an appropriate sized cable used. Consideration must be given to thermal effects, grouping and ambient temperature.

Cooker circuit Domestic cookers are manufactured in a wide range of power outputs. It is important that the cable selected is suitable for the current demand and where it is to be placed; again taking into consideration any correction factors which might have to be applied. An allowance can be made for diversity of power consumed, otherwise a cable of enormous size would result. In general, for an individual household, the following expression can be used:

$$I = 10 + (0.3 \times I_t) \qquad [6.9]$$

Should a socket outlet be incorporated within the cooker control switch, then:

$$I = 15 + (0.3 \times I_t) \qquad [6.10]$$

where I is the current in amps after an allowance
for diversity has been made and
I_t is the total current demand of the
appliance.

As a practical example, consider a standard cooker
having a total load of 10.7 kW (10 700 watts).
This is connected to a 230 V supply, served by a
cooker control switch incorporating a socket outlet.
By using Expression [2.5], the maximum current
demand may be found:

$$I = \frac{W}{V}$$

Substituting figures,

$$I = \frac{10\ 700}{230}$$

$$I = 46.521\ \text{amps}$$

Expression [6.10] allows for a diversity to be
made:

$$I = 15 + (0.3 I_t)$$

Substituting figures,

$$I = 15 + (0.3 \times 46.5)$$
$$I = 28.95\ \text{amps}$$

In practical terms the cable required to serve the
cooker would be rated higher than 29 A. A
6.0 mm^2 PVC insulated and sheathed cable would
suit the requirements of this installation but
environmental correction factors must be taken into
consideration when selecting the size of cable
required. (An example may be seen in Appendix
C).

In order to satisfy the demands of Regulation
476-03-04, a suitably rated switch must be
provided as means of isolation and placed as near
as practical to the appliance.

Figure 6.28 illustrates the standard method of
wiring to a cooker appliance where the hob and
oven are separate from each other. Potentially both
appliances have a high current demand and are
served by a common cable whose current-carrying
capacity is less, but the circuit is protected by a 30
or 32 A overcurrent device. When a cooker hood
is required, the supply may be spurred from the

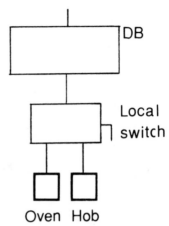

Figure 6.28. Basic wiring requirements for a domestic cooker
circuit.

kitchen final ring circuit, terminating into a fused
connection unit. Alternatively, an independent
circuit could be installed from the local distribution
board using a suitably sized lighting grade cable
terminating in a 20 A double pole switch. A semi-
enclosed fuse (rewireable) to BS 3036 or circuit
breaker may be used as overcurrent protection and
should be rated at no more than 5 A.

Fixed wall heater circuit Fixed wall heaters are
becoming more popular now and are usually rated
at between 1 and 3 kW.

Until recently, it was customary to wire smaller
types of infrared heaters, rated at 0.75 kW,
directly into the permanently live side of the local
lighting circuit. Now higher wattage heaters are
available it is essential that a separate and
independent circuit is installed to accommodate the
extra load. Figure 6.29 shows an example of this
type of circuit, terminating with a double pole
switch. Should the appliance be installed in a
bathroom, the local isolating switch would then be
replaced by a suitable blanked face connection
unit; control being effected by means of an
integral pull switch within the heater. Regulation
476-03-04 confirms.

Installation cable size, serving the appliance,
may be checked by referring to Regulation Table
4D2A once current demand and correction factors
have been calculated. It is usual to wire using
2.5 mm^2 PVC insulated and sheathed cable but if
in any doubt, check the power output in watts of
the heater and if necessary use a larger sized
cable. It is far better to be safe than sorry.

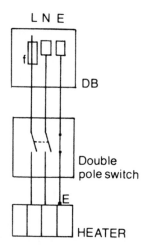

Figure 6.29. Basic wiring requirements for a fixed wall heater. When the installation is designed to be incorporated within a bathroom, the isolating switch must be fitted outside and adjacent to the bathroom door.

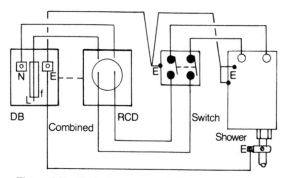

Figure 6.30. The basic requirements for an electric shower circuit incorporating a residual current device which may be added for additional protection.

Electric shower circuit Electric shower units are obtainable in various ratings. Once the power output in watts has been established, Expression [2.5] may be applied in order to calculate the potential current drawn from the supply. Consideration must also be given to thermal effects, grouping and ambient temperature and whether a semi-rewireable fuse be incorporated within the circuit. The *Wiring Regulations* provide an excellent guide for selecting the size cable required for the task in hand. Usually the cable is the same as if installing an average sized domestic cooker. However, it is preferable to check the total loading as the appliance may require a larger conductor. Figure 6.30 illustrates the basic circuit requirements. It is well worth remembering that the shower must be permanently connected to the electrical supply by means of a 30 or 40 A double pole switch which has a mechanical switching movement separated by at least 3 mm. The switch must be clearly identifiable and would normally take the form of either a pull switch mounted on the ceiling or a double pole wall switch sited *outside* the shower room. Wiring must originate directly from the control switch without the use of a plug and socket arrangement.

It is important to note that an independent current protective conductor is installed to serve as a supplementary equipotential bond between simultaneously accessible exposed conductive parts of equipment and extraneous metal forming part of the infrastructure of the shower room.

For additional safety, a 30 mA residual current device may be incorporated in circuit to provide automatic disconnection facilities should a fault condition arise. Regulations 554-05-02 and 03 confirm.

Door bell circuit A door bell installation may either be connected to an appropriate size battery or alternatively from a bell transformer. Since current demand is very low, a transformer may be connected to the permanently live side of a convenient lighting circuit. Should this be the chosen method, it is preferable to wire the supply by means of a fused connection unit before serving the transformer. The size of the local cartridge fuse should be reviewed and, if necessary, changed to suit the rating of the transformer.

If this method is unsuitable, the bell transformer can be supplied directly from the distribution centre connected to a suitable overcurrent device. The circuit is then terminated with a double pole switch. The switch allows electrical isolation should it be required to turn off the circuit for any reason.

The low voltage side of the transformer may be wired using standard twin bell wire. It is important to take into account the effect of volt drop should both bell and push be placed at some considerable distance from the transformer. This phenomenon is far more apparent when the installation is reliant upon batteries as means of an energy source.

As a practical example, consider the following problem:

A 3 V combined bell and battery unit recently installed to serve a large warehouse sounded feeble even though the terminal voltage was found to be 3.0 V. The current drawn from the circuit was measured and found to be just 0.45 A whilst the total resistance of the circuit was calculated to be 3.2 Ω. Prove theoretically why the bell sounded so weak.

By applying Expression [2.7]:

$$\text{volt drop} = I \times R$$

Substituting figures for known values,

$$\text{volt drop} = 0.45 \times 3.2$$
$$\text{volt drop} = 1.44 \text{ volts}$$

(and also the answer to why the bell sounded so feeble).

In practical terms it would have been far more appropriate to install a suitable stepdown transformer with a 5 or 6 V tapping to serve the installation. Figure 6.31 shows details of how a bell circuit may be wired using a simple stepdown transformer. Basic theoretical aspects of the transformer are reviewed in Chapter 7.

Garage installation The criteria determining the size and type of supply cable serving a home-based garage may be drawn from the expected current demand, the route by which the cable is taken and any conductor correction factors (Appendix C) which might need to be taken into consideration.

Once the total power consumption has been evaluated, Expression [2.5] may then be applied in order to calculate the maximum potential current drawn from the supply. A suitable sized cable could then be chosen to meet the requirements of the installation.

Assuming the garage is detached from the house, the service cable may either be routed overhead by means of a catenary wire or placed underground. If an underground route is preferred, a PVC insulated, stranded wire armoured cable would be the most appropriate to use. Should the installation be under financial constraints, an overhead route might be wiser. For additional safety, both the garage installation and service cable can be protected by a residual current device with a residual operating current of 30 mA. Figure 6.32 describes in schematic form how an installation of this type might be installed.

As a practical example, consider the following:

A recently built garage was designed to include the following electrical circuits:

1. *One 3 kW immersion heater.*
2. *One 100 W lighting point.*
3. *One fused connection unit serving a fixed 3 kW wall-fire heater.*

Calculate the total potential current demand on the proposed installation (230 V single phase supply).

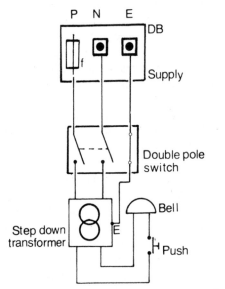

Figure 6.31. Simple bell circuit using a step down transformer.

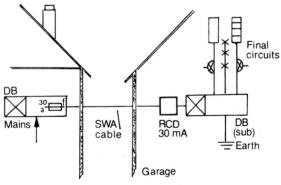

Figure 6.32. A typical garage installation and general requirements.

Total potential power
= 3000 + 100 + 3000 watts
Total potential power
= 6100 watts

The potential current may now be calculated by applying Expression [2.5]:

$$I = \frac{W}{V}$$

Substituting figures,

$$I = \frac{6100}{230}$$

$$I = 26.5 \text{ amps}$$

By referring to the *Wiring Regulations*, a suitably sized cable may be selected, taking into account volt drop, type of cable to be used and correction factors. (An example of correction factors may be seen in Appendix C.)

Off-peak, or restricted use, installations Off-peak installations are designed to provide electricity at a cheaper rate to the consumer. The general requirements are outlined in Figure 6.33. Both credit meter and time-switch are property of the local electricity supply authority and although part of the installation, play no financial role to the electrician where initial costing is concerned.

A suitable sized distribution centre is needed to accommodate a system of radial circuits terminating in 20 A double pole switches. This provides a means of isolation for each appliance fitted. Domestic installations are connected to loads where an advantage may be gained using cheaper off-peak electricity to power night storage heating equipment and immersion heaters. Agricultural

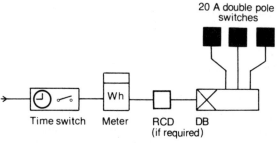

Figure 6.33. General wiring requirements for an 'off-peak' installation.

establishments take advantage of this service to run refrigeration plants in order to produce inexpensive ice for bulk milk tank installations.

Domestically, the current demand for the majority of off-peak circuits is no more than 16 A. Both night storage and immersion heater circuits, up to a 3 or 4 kW loading, are usually wired using 2.5 mm^2 cable. It is important to remember to wire only one appliance per circuit when carrying out an off-peak electrical installation, otherwise an overcurrent could arise. Never wire a final ring circuit to serve storage heaters. The current demand could be far too great to support the 30 A overcurrent protection device used to protect the circuit. Always wire in radial circuits with one circuit for each appliance.

Industrial circuit arrangements
Figure 6.34 illustrates a typical wiring method used to control an industrial compressor motor serving a cold room installation. Each control switch is wired in series formation and provides a means of de-energising an electromagnetic coil to release a contactor switching assembly. Other than the control circuit isolation switch, shown in Figure 6.34, all other switching devices are automatic and controlled by outside forces such as temperature, high and low pressure or time clocks. These are shown in Figure 6.34 as HP and LP, representing the high and low pressure switches and t^0, the control thermostat.

During the working cycle all self-activating and manually operated switches are closed, enabling the coil circuit to be energised. This action brings the connected load into play and the machine will operate. A control device automatically switching to an open mode will immediately de-energise the coil circuit and the machine will stop.

It is important to remember to fuse the control circuit separately with an overcurrent protection device rated to the current demand of the coil. Should the coil circuit be supplied, unprotected, from the unswitched side of the contactor, the overcurrent protection will be excessively high and no safeguard can therefore be afforded to the coil.

The control circuit concept can be applied to any type of circuit requiring external supervision. Installations such as these are usually carried out

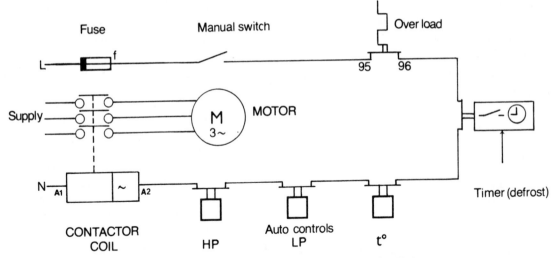

Figure 6.34. A basic control circuit serving a coldroom installation.

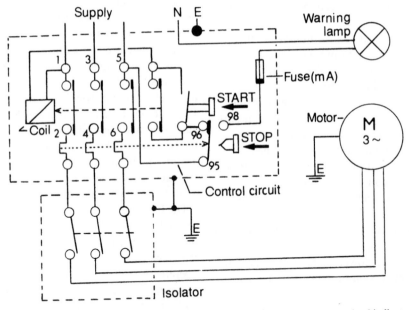

Figure 6.35. Direct on-line motor control circuit featuring a supplementary remote overload indicator facility

using solid drawn steel or plastic conduit and wired throughout with PVC insulated cables. As with other installations, an appropriate sized cable must be chosen to supply the load. Control circuit wiring may be installed using a smaller conductor, as current demand will be far less than the connected load.

Motor circuit with remote overcurrent warning facility A simple machine circuit with a remote overcurrent warning facility may be wired as shown in Figure 6.35. The rating details, or the amount of current the motor may safely draw from the supply before a critical overcurrent situation occurs, may normally be found on the manufacturer's nameplate. The value obtained is an unspecific guide to the assumed cable size. Although less academic than the theoretical approach it serves as well in times of emergency; providing the physical length of the installation is short.

Warning indicator circuit Remote overcurrent warning indicator circuits are wired using a lighting grade cable as current demand is extremely small. Provision must be made for local overcurrent protection which is independent from the motor circuit.

Volt drop: selecting the size of cable required

Wiring Regulation 525-01-02 is satisfied when a voltage drop occurring between the supply authority's terminal and the fixed current using equipment does not exceed 4 per cent of the nominal voltage (U_0) of the supply.

A drop in voltage which is excessive will have a direct effect on other current-using equipment. This must be taken into consideration when selecting the size of cable required.

As an example, consider the following. Correction factors may be ignored for this example:

A 20 m, 230 V, single phase circuit designed to supply 27.75 A is to be wired in single copper PVC insulated cable and drawn through welded steel conduit. Calculate the minimum size cable which may be used to comply with the electrical regulations governing volt drop.

Given that:

maximum volt drop permitted
= 0.04 × voltage   [6.11]

Substituting for figures,

maximum volt drop permitted
= 0.04 × 230
maximum volt drop permitted
= 9.2 volts

Given that:

$$\text{Actual volt drop} = \frac{\text{mV/A/m} \times I \times l}{1000}$$
[6.12]

where mV/A/m is the millivolt drop per amp per metre,

I is the current flowing in the conductor and

l is the length of the conductor in metres.

Substituting for known values:

$$9.2 = \frac{\text{mV/A/m} \times 27.75 \times 20}{1000}$$

Cross-multiplying,

$$9.2 \times 1000 = \text{mV/A/m} \times 555$$

and dividing each side of the equation by 555,

$$\text{mV/A/m} = \frac{9200}{555}$$

$$= 16.57 \text{ mV/A/m}$$

Choice of size
By reviewing Regulation Table 4D2B a suitably sized conductor can be selected having a volt drop of less than 16.57 mV/A/m (Figure 6.36). To satisfy this criterion academically a minimum conductor size of 6.0 mm² must first be considered having a volt drop of 7.3 mV/A/m.

This is an ideal method to use where long lengths of cable are employed but consideration should be given to the ambient temperature if a more accurate evaluation of volt drop is required.

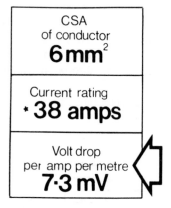

Figure 6.36. Volt drop per amp per metre and current-carrying capacity for a 6 mm² copper conductor. (* Installed in conduit; two cables; single phase AC.)

Fibre optics: a brief insight

In 1977 Britain was able to present to the world the first commercial fibre optic telephone link. Optical-fibre technology has since grown and many long-distance telecommunication networks have been established using glass fibre-optic cables.

Many of these long-distance cables are now combined within the earth/lightning conductor serving the *National Grid System*. Changing the original earth conductor for a combination earth/fibre-optic cable was both time consuming, expensive and entirely dependent on sophisticated machinery. Once commissioned the installation was able to provide total security and avoided hazards and damage associated with cable management systems of a more conventional type.

An optical-fibre communication cable employed for data, telephone channels or television programmes is very small, surprisingly only some 0.125 mm in diameter, Figure 6.37. It supports a thin glass core which, compared to the cladding surrounding, has a much higher *refractive index*. (This may be defined as the ratio of the *Sine* of the angle of incidence to that of the subsequent angle of refraction when a light source passes from one medium to another. Glass, for example, has a refractive index of between 1.5 and 1.7.) Either an *infra-red* light source just below the human visible spectrum or a low-powered LASER is used as an optical transmitter. As the central glass core has a much higher refractive index than its surrounding cladding, the light source is totally imprisoned and rebounds off the junction of the core and cladding by internal reflection. This process allows the infra-red source to penetrate through the glass core at a great speed. Visible light is not favoured as its intensity reduces due to absorption within the glass core.

Figure 6.38 depicts a block diagram of a typical fibre-optic telecommunication system. Speech, converted into electrical signals, is first pulse code modulated in a *coder* and then fed into an *optical transmitter* where signals are changed to a light source and projected through the glass core of the fibre-optic cable. At the point of termination an *optical receiver* converts the infra-red light source back into electrical pulses which are then *decoded* into speech or data.

Light sources

Light for optical transmission can be derived from the following sources:

1. *Light amplification by the stimulated emission of radiation.* (LASER).
   The type of LASER used for fibre-optic installations is made from *gallium arsenide phosphide*. Although far more expensive, it can be switched on and off by pulses of monitored current far more quickly than if a light emitting diode were in its place.
2. *Light emitting diode.* (LED)
   This is a far cheaper way of providing for a light source. Light emitting diodes are manufactured from the semi-conducting material *gallium arsenide phosphide*. When used as an optical transmitter the device emits infra-red radiation at a wavelength of approximately 0.84 micro metres. (1.0 micro metre is one millionth part of a metre).

Advantages of using a fibre-optic communication system

1. Free from electrical interference.
2. Far lighter and easier to handle than copper cable.
3. User security greatly improved.
4. Volt drop is not applicable.

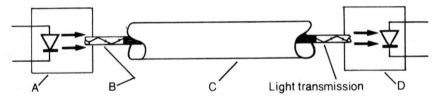

Figure 6.37. A fibre optic cable. (A, guided light transmitter; B, glass core; C, cladding; D, guided light receiver).

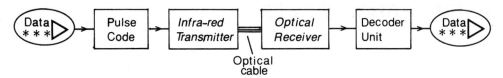

Figure 6.38. A fibre-optic communication system.

5. The system is able to span far greater distances than conventional wiring methods.
6. Fibre-optical networks are able to carry far more information than by conventional means. To date: two colour television channels or up to 2000 telephone channels.

Summary

In this chapter we have reviewed a brief history of cables and methods of installation and demonstrated the effect of volt drop due to temperature variations.

Factors such as cross-sectional area, length and the material composition of the conductor were shown to have an influence on both resistance and current flow. Materials such as copper, silver and aluminium were all shown to have positive temperature coefficients, whereas carbon and silicon display negative features.

Ways of calculating resistance were explored both theoretically and practically and it was found that the resistance of a cable was directly proportional to length but inversely proportional to cross-sectional area.

Power and final circuit arrangements were examined to provide both a practical and theoretical insight into installation problems. Methods of testing were also analysed.

In Chapter 7 we will be investigating the basic principles of the transformer and its relationship between the fundamental concepts studied in previous chapters.

7 Basic transformers

In this chapter: Brief history. Basic fundamental principles and practical applications. Transformer calculations and testing.

On 29 August 1831 *Michael Faraday* arrived at his greatest discovery. He successfully demonstrated that a transient electric current in coil wound round an iron ring would cause a momentary current in another coil wound adjacent to it. This experiment was no more than a scientific curiosity and the full implication was not realised until much later.

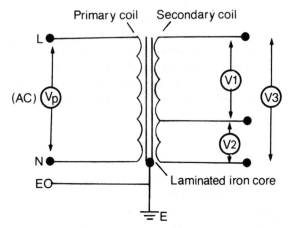

Figure 7.1. A simple transformer.

Fundamental principles

The transformer is one of the simplest devices used in electrical engineering today. It has no mechanical moving parts and can be made to any size to accommodate any task. It comprises two insulated coils of wire called the *primary* and *secondary* windings, wound separately from each other on common laminated iron core, (Figure 7.1).

The transformer works on the principle of the rise and fall of a magnetic force field produced by an alternating voltage applied to the primary winding. This causes a fluctuating magnetic field in the primary coil and by mutual induction, an alternating current is induced into the secondary coil. A similar effect can be demonstrated when a simple loop of wire is rotated within a fixed magnetic field. As the loop cuts through the lines of flux at right angles, an emf is induced into the loop.

Types of transformer

If both primary and secondary windings have the same number of turns of wire, the voltage induced in the secondary would be equal to the applied primary voltage. If the turns forming both windings are not the same, then the voltages will be proportional, but different.

A transformer with a primary coil of 100 turns and a secondary of 1000 turns and a supply of 10 V applied across the primary winding will induce 100 V into the secondary coil. A transformer will only function when an alternating voltage is applied to its primary coil. Should direct current be administered, the transformer would act as an electromagnet or possibly burn out.

The frequency of the supply is completely unchanged; only the secondary voltage is either increased, decreased or stays the same. If the transformer supplies a higher voltage than the voltage applied, it is called a *step up transformer*. If a lower voltage is obtainable, then it is known as a *step down transformer*. When the same number of turns apply for both primary and secondary windings, the device is known as an *isolating transformer*.

Often a ratio is applied, as for an example 1 : 4. This implies that for every one primary turn there are four secondary turns. Transformers are very

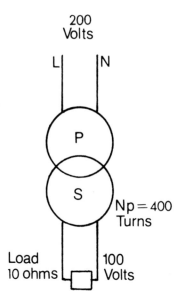

200
Volts

L N

P

S

Np = 400
Turns

Load
10 ohms

100
Volts

Figure 7.2. Schematic arrangement of a two to one step down transformer.

$$\frac{E_p}{E_s} = \frac{I_s}{I_p} \qquad [7.1]$$

where E_p is the voltage applied to the primary winding,

E_s is the voltage induced into the secondary winding,

I_p is the current, in amps, in the primary winding and

I_s is the current flowing in the secondary winding.

Substituting for figures:

$$\frac{200}{100} = \frac{10}{I_p}$$

Cross-multiplying and evaluating in terms of I_p:

$$I_p = \frac{1000}{200}$$

$$I_p = 5 \text{ amps}$$

efficient machines, having an average capability of approximately 96 per cent.

Calculating current drawn Figure 7.2 outlines in schematic form a diagram of a 2 : 1 step down transformer supporting a 400 turn primary winding and served by a 200 V AC supply. The secondary winding produces 100 V carrying a load with a resistance of 10 Ω.

By applying Ohm's Law, Expression [2.1], the current drawn from the secondary winding may be calculated:

$$I = \frac{E}{R}$$

where E is the secondary AC voltage.

Substituting figures,

$$I = \frac{100}{10}$$

$$I = 10 \text{ amps}$$

Once the current flowing in the secondary winding has been found, the following expression can be used to calculate the current drawn from the primary winding:

Calculating power taken from the circuit This might, at first glance, seem as though we are getting something for nothing, but by applying Expressions [7.2] and [7.3] and finding the total power in both primary and secondary windings it may be clearly seen that we are not:

Primary power $= I_p \times E_p$ [7.2]
Primary power $= 5 \times 200$
Primary power $= 1000 \text{ watts}$

Secondary power $= I_s \times E_s$ [7.3]
Secondary power $= 10 \times 100$
Secondary power $= 1000 \text{ watts}$

Calculating the number of turns of wire In order to calculate the number of turns in the secondary winding, the following expression must be used:

$$\frac{E_p}{E_s} = \frac{N_p}{N_s} \qquad [7.4]$$

where N_p is the number of turns in the primary winding and

N_s is the number of turns in the secondary winding.

As an example consider Figure 7.2, illustrating

that N_p = 400 turns. Substituting for figures given or obtained:

$$\frac{200}{100} = \frac{400}{N_s}$$

$$200N_s = 400 \times 100$$

$$N_s = \frac{40\,000}{200}$$

$$N_s = 200 \text{ turns}$$

Calculations have shown that the transformer has a ratio of 4 : 2 or 2 : 1.

Checking A way to check the calculation is by expressing the ratio in terms of current drawn from both primary and secondary windings.

$$\frac{N_p}{N_s} = \frac{I_s}{I_p} \qquad [7.5]$$

where N_p is the number of turns in the primary winding,

N_s is the number of turns in the secondary winding

I_p is the current flowing in the primary winding and

I_s is the current drawn from the secondary winding.

Substituting figures and evaluating in terms of I_p:

$$\frac{400}{200} = \frac{10}{I_p}$$

$$400I_p = 200 \times 10$$

$$I_p = \frac{2000}{400}$$

∴ Primary current I_p = 5 amps

Calculating the value of the secondary current I_s and substituting figures,

$$\frac{400}{200} = \frac{I_s}{5}$$

$$200I_s = 400 \times 5$$

$$I_s = \frac{2000}{200}$$

∴ Secondary winding current I_s = 10 amps

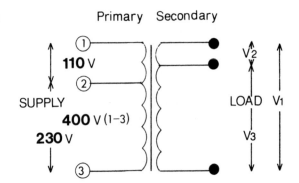

Figure 7.3. A multivoltage transformer.

By Calculation ($I \times V$), it can be shown that both primary and secondary windings generate the same amount of power.

Multivoltage transformer

Figure 7.3 schematically illustrates a multivoltage transformer which basically is constructed in a similar fashion to the standard type already debated on these pages. By studying Figure 7.3 it can be seen that provision has been made for several voltage ranges to serve both primary and secondary windings.

Testing

Testing a transformer may be carried out by use of a voltmeter and a suitable load. Alternatively, a standard continuity meter may be employed by measuring the resistance values of both primary and secondary windings. However, when adopting this method ensure that the transformer is completely isolated from the supply. Also, remember that both size and number of turns of wire will produce different values for each winding. Should the transformer under test be an isolating transformer, then both sets of windings will in effect produce similar values.

Although open circuit conditions are reasonably straightforward to detect, short circuiting due to a breakdown of insulation is far more difficult to assess unless prior knowledge has been gained of the expected value in ohms of both sets of windings.

Voltage tests taken without a connected load will

probably produce a voltage slightly higher than expected.

Summary

The transformer is completely reliant on the principle of an alternating magnetic flux to enable an electromotive force to be induced into a secondary winding. At first sight, it appears that something is being gained for nothing as the induced voltage can often be higher than the voltage administered. By calculation it was proved that the power generated in both sets of windings was equal. Hence:

$$E_p \times I_p = E_s \times I_s$$

Under the aegis of Ohm's Law, the current drawn from the secondary side of the transformer may be calculated. Power developed was determined by use of Expressions [7.2] and [7.3] ($W = I_p \times E_p$ or $W = I_s \times E_s$).

Testing was carried out by means of volt and continuity meters to determine both voltage output from the secondary winding and the resistance in ohms offered to the supply from both sets of windings.

Transformers are constructed to suit the needs of industry. They can be made extremely small, as with a 'one-to-one' isolating transformer used in radio and television circuitry, or designed large enough to accommodate the demands of the National Grid. Whatever size and purpose, all share the same basic fundamental principles discovered by *Michael Faraday* in 1831. The role they play in electrical engineering is of paramount importance to the industry they serve.

Part 1 review questions

Listed are 25 questions to test your new-found basic knowledge and understanding of electrical theory. The answers are given in Appendix A.

1. Name the three principal parts of an atom.
2. How many times heavier is a proton compared to the electron it accompanies?

   (a) 1845
   (b) 1830
   (c) 18×10^2
   (d) 1840

3. Resistors of 6, 4 and 3 Ω are connected in parallel formation. Calculate the total resistance offered in ohms.
4. What expression may be used to solve problems relating to the resistance of a conductor due to variations in temperature?
5. Identify the two best conductors of electricity from the list provided:

   (a) Silicon
   (b) Copper
   (c) Carbon
   (d) Silver
   (e) Nickel
   (f) Mercury
   (g) Aluminium

6. For what reason is the valency shell of an atom important for electrical conductivity?
7. Listed are four expressions, only one of which is correct, to calculate the power developed in a circuit when only the voltage and resistance are known. Which is correct?

   (a) $W = R/V^2$
   (b) $W = V^2/R$
   (c) $W = R \times V$
   (d) $W = \sqrt{V} \times R^2$

8. Describe how the internal resistance of a battery may be calculated.

9. Calculate the total resistance in ohms of 6, 4 and 3 Ω resistors connected in series formation.
10. Specify two reasons, other than fault conditions, why nuisance tripping would occur to a circuit served by a residual current device.
11. State the reason why a voltage-operated earth leakage circuit breaker would require an earth electrode.
12. Name two types of direct current electric motors.
13. State three categories of alternating current electric motors.
14. Listed are four expressions, only one of which is correct, to calculate the value of capacitors wired in series formation. Which is correct?

   (a) $\dfrac{1}{C_t} = \dfrac{C_1}{1} + \dfrac{C_2}{1}$

   (b) $\dfrac{1}{C_t} = \dfrac{1}{C_1} + \dfrac{1}{C_2}$

   (c) $C_t = C_1 + C_2$

   (d) $\dfrac{1}{C_t} = C_1 + C_2$

15. Describe a common use for an AC shaded pole single phase electric motor.
16. Define a common fault condition occurring in capacitors.
17. Explain the term *capacitative reactance*.
18. Specify four factors affecting the resistance of a conductor.
19. What is meant by the term *negative coefficient of resistance*?
20. Name the three principal parts of a transformer.
21. Listed are four expressions, only one of which is correct, which allows for the diversity of an

electric cooker to be calculated. Which is correct?

(a) $I = 15 + (0.4I_l)$
(b) $I = (15I_l) + 0.4$
(c) $I = 15 + (0.3I_l)$
(d) $I = (I_l15) + 0.3$

22. Why is silver a better conductor of electricity than nickel?

23. Find the correct function of an isolating transformer from the following:

(a) Used when AC and DC supplies require segregation from each other.

(b) Used in radio and television circuits to isolate different frequencies.

(c) A transformer designed with both primary and secondary windings having an equal number of turns.

(d) A device designed in order to reduce the incoming supply voltage.

24. Given the value of both primary and secondary voltages and the connected load in ohms, how may the current flowing in the primary winding be calculated?

25. Explain the term *electrical resistivity*.

A summary of theoretical expressions used in Part 1

[1.1]   electrons per quantum $= 2N^2$

[1.2] $I = \dfrac{V}{R}$

[1.3] $Q = I \times t$

[1.4]   resistance $= \dfrac{\text{voltage}}{\text{current}}$

[1.5]   power $=$ current $\times$ voltage

[2.1]   current in amps $= \dfrac{\text{voltage}}{\text{resistance}}$

[2.2]   power in watts $= \dfrac{V^2}{R}$

[2.3] $I = \dfrac{W \times 1.8}{V}$

[2.4] $W = IV$

[2.5] $I = \dfrac{W}{V}$

[2.6] $R_t = R_1 + R_2 + R_3$

[2.7]   volt drop $I \times R$

[2.8]   Voltage applied $= VD_1 + VD_2 + VD_3 + VD_4$

[2.9] $W = I^2R$

[2.10] $R_t = R_f + R_1$

[2.11] volt drop $= I \times R_f$

[2.12] $I = \dfrac{V}{R + r}$

[2.13] $V = I(R + r)$

[2.14] $r = \dfrac{V}{I} - R$

[2.15] $r = \dfrac{E - V}{I}$

[2.16] $I = \dfrac{V}{R + r_1 + r_2 + r_3 + r_4 + r_5 + r_6}$

[2.17] $I = \dfrac{V}{R + (r/2)}$

[2.18] $\dfrac{1}{R_t} = \dfrac{1}{R_1} + \dfrac{1}{R_2}$

[2.19] $R_t = \dfrac{R_1 \times R_2}{R_1 + R_2}$

[4.1] $I_a = \dfrac{V - E}{R_a}$

[4.2] $I_a \times R_a = V - E$

[4.3] $V = E + (I_a \times R_a)$

[4.4] $E = V - (I_a \times R_a)$

[4.5]   slip % $= \dfrac{N - \text{rotor speed}}{N} \times \dfrac{100}{1}$

[4.6] $N = \dfrac{f}{P}$

[4.7]   power factor $= \dfrac{W}{VA}$

[4.8]   reactive power $= \sqrt{(VA^2 - W^2)}$

[4.9] $X_C = \dfrac{E^2}{\text{reactive power}}$

[4.10] $X_C = \dfrac{1}{2\pi fC}$

[4.11] $C = \dfrac{1}{2\pi fX_C}$

[5.1] $I = \dfrac{E}{X_C}$

[5.2] $\mu F = \dfrac{3182.6801 \times \text{current}}{\text{voltage}}$

[5.3] $I = \dfrac{\mu F \times \text{voltage}}{3182.6801}$

[5.4] $\text{Voltage} = \dfrac{3182.6801 \times \text{current}}{\mu F}$

[5.5] $\dfrac{1}{C_t} = \dfrac{1}{C_1} + \dfrac{1}{C_2}$

[5.6] $C_t = \dfrac{C_1 \times C_2}{C_1 + C_2}$

[5.7] $C_t = C_1 + C_2$

[5.8] $\dfrac{1}{C_t} = \dfrac{1}{C_1} + \dfrac{1}{C_x}$

[5.9] $Z = \sqrt{(R^2 + X_C^2)}$

[5.10] $X_L = 2\pi f L$

[5.11] $I = \dfrac{E}{X_L}$

[5.12] $I = \dfrac{E}{(2\pi f L)}$

[5.13] $Z = \sqrt{[R^2 + (X_L - X_C)^2]}$

[5.14] $I = \dfrac{E}{Z}$

[6.1] $R = \dfrac{\rho l}{a}$

[6.2] volt drop $= I\left(\dfrac{\rho l}{a}\right)$

[6.3] $\dfrac{R_1}{R_2} = \dfrac{1 + (\alpha t_1)}{1 + (\alpha t_2)}$

[6.4] $°C = \dfrac{(°F - 32) \times 5}{9}$

[6.5] $°C = K - 273.15$

[6.6] permissible volt drop $= U_0 \times 0.04$

[6.7] $R \simeq \dfrac{1}{2} \times \dfrac{\text{1st step reading}}{1}$

[6.8] $R \simeq \dfrac{\text{1st step reading}}{4} + \dfrac{\text{2nd step reading}}{4}$

[6.9] $I = 10 + (0.3 \times I_t)$

[6.10] $I = 15 + (0.3 \times I_t)$

[6.11] maximum permitted volt drop
$= 0.04 \times U_0$

[6.12] actual volt drop $=$

$\dfrac{\text{mV/A/m} \times I \times \text{length (m)}}{1000}$

[7.1] $\dfrac{E_p}{E_s} = \dfrac{N_p}{I_p}$

[7.2] power in watts in the primary $= I_p \times E_p$

[7.3] power in watts in the secondary $= I_s \times E_s$

[7.4] $\dfrac{E_p}{E_s} = \dfrac{N_p}{N_s}$

[7.5] $\dfrac{N_p}{N_s} = \dfrac{I_s}{I_p}$

For a list of abbreviations please turn to p.xi at the front of the book.

Part 2 Practical

Part 1 of this book was fashioned to show the theoretical concepts of electrical installation work. In Part 2 the emphasis will be placed on the practical side of the skill, coupled with methods and techniques needed to achieve a working knowledge of the task to be undertaken.

Starting a major electrical installation is a straightforward operation provided the job is thought through before any work is carried out. Rushing into a job without sufficient planning could lead to chaos at a later stage. Take time to study the best and most efficient way of carrying out the installation. Check the merits and disadvantages of your chosen system of wiring. This is important when, for example, lighting circuits are installed, as there are many different techniques and methods which may be adopted. It might be prudent not to employ the joint box method of wiring if, for example, a tiled floor is to be laid throughout. Imagine the inconvenience caused should an installation fault be discovered after the tiled floor had been laid, let alone the aggravation generated by other tradespeople.

Avoid rashness, weigh the consequences of any chosen path and choose the correct wiring method to suit the job. This way installation problems will be minimised.

If practical, produce a deadline when work should be finished and look ahead for productivity targets. It is easy to slip behind through inefficiency! A planned job within a well-structured framework should be your main objective. Decide on your aims before work commences; this will enable the task to be carried out far more smoothly.

Try to maintain a good standard of workmanship. At times it can be tempting to fall below your personal expectations. It is easy to say, 'Oh, that will do!' when a little more time and precision targeted at the job in hand would produce far better results. Often there are pressures outside our field of influence which dictate this approach, such as production costs and finishing dates. It is far better to take a little longer and get it right first time, than to hurry and overstretch your capabilities, only to have to return to put it right!

Working in a customer's home can have unforeseen consequences should the installation be treated in a similar manner as an industrial site. Clients are often very sensitive and inwardly resent the intrusion into their home by strangers. Always advise your customer of your intended programme of work. Rewiring can cause much stress and strain on relationships. Never attempt to move, resite or push aside anything of value in order to carry out your programme of work. Imagine, for example, a small table cluttered with personal artefacts, all of which are either valuable or highly sentimental. One clumsy or awkward move could spell disaster should one of these precious items crash to the floor. The golden rule must always be: always ask your customer to remove or make safe any items which might be of value in the area in which you are to work. Remember: Should anything be discovered broken, damaged or soiled, you will be blamed, whether it was your responsibility or not.

Customers are usually quick to claim compensation for damage inadvertently caused by electricians whilst undertaking domestic installation work. If work is carried out in a client's residence, remember these simple guidelines:

1. Carry out the job in a professional manner.
2. Trust no one but yourself.
3. Always maintain a business relationship with your customer.
4. Acquaint the customer with the intended programme of work.
5. Leave the job clean and tidy.

The day to day running of an electrical

installation serving a large construction site can often become quite involved. Never commit anything important to memory; keep a day diary. List names, times, extras, variations and verbal instructions given by other trades and the general foreman or site manager.

Always keep a spare copy of the current electrical drawings in a safe place in case the working drawings are destroyed or damaged. If you are confused, never be afraid to ask. It is far better to get things right first time than to guess wrongly. It might cost you your job!

Manufacturers' drawings, information sheets and technical handouts are all very useful items to keep. Reference may be made to them at any time in order to gain or recall information. They are best kept in a folder or suitable box, and make a useful addition to any tool kit.

Part 2 provides an ideal introductory insight into electrical installation techniques and problems. It will enable the electrical worker to build on existing knowledge and skills, thus leading to a more sound understanding of the task to be undertaken.

This section has been written to meet with the requirements of the revised version of the *City and Guilds GCL1 236 Electrical Installation Syllabus*, the *National Vocational Standards* for the electrical engineering industry and the *16th Edition of the IEE Wiring Regulations*.

8 Electrical supply systems and bonding techniques

In this chapter: Electrical supply systems. Bonding techniques. Special requirements for bathrooms. Supplementary bonding. Bonding to radiators. Protective multiple earthing systems. Current protective conductors. Disconnection times. Earth-free bonding.

Types of distribution systems: earthing arrangements

In Britain electricity generating companies have to comply with the *1937 Electrical Supply Regulations*. These require each supply network to be earthed at one point. There are many different types of distribution systems to be found throughout the country; however, Regulation 312-03-01 provides for the following categories: TT system, TN-S system, TN-CS system, TN-C system and IT system. IT systems are not used for public supplies.

All supply and distribution systems are catalogued using letters of the Roman alphabet. The first letter indicates the *earthing arrangements at the supply source*. In the majority of cases this will refer to the local distribution transformer. The second letter suggests the *installation earthing arrangements* and the third letter points to the *earthed supply conductor arrangements*.

Reviewing the initials

The first letter

T (*terra firma*): At least one point of the supply is connected directly to earth. Figure 8.1 illustrates.

The second letter The second letter points to the

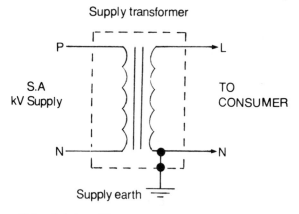

Figure 8.1. In a 'TT' system the neutral side of the supply to the consumer is connected directly to earth at the supply transformer.

relationship between the exposed conductive parts and earth. Figures 8.2 and 8.3 will clarify.

T: conductive parts connected directly to earth and independent of the supply earth.

N: conductive parts connected directly to the earth point at the source of the supply. By definition, this can refer to the metal infrastructure in which the transformer is housed.

The third letter The third and final letter indicates the arrangement of the protective and neutral conductor serving a TN system.

S: both protective and neutral conductors are separate from each other.

C: both protective and neutral conductors are combined together in one cable. This could take the form of a single or multicored PEN cable.

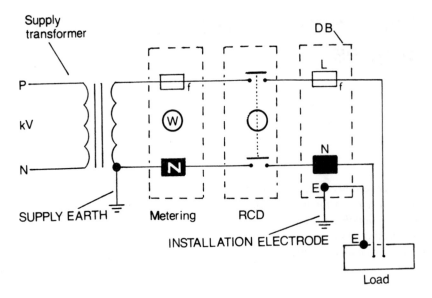

Figure 8.2. In a 'TT' system conductive parts are connected directly to earth and are independent of the supply earth.

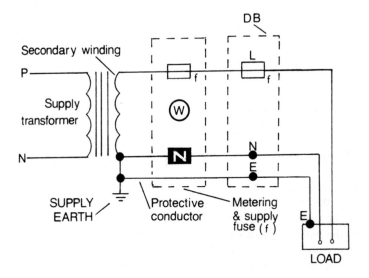

Figure 8.3. In a 'TN-S' system the conductive parts are connected directly to the earth point at the supply transformer.

Power distribution systems

TT systems

The supply of electricity to rural Britain is usually delivered by means of a network of overhead cables ministered from a neighbourhood transformer. This method of power distribution is known as a TT system and effectively provides a phase and neutral supply to the consumer. The customer is then responsible for the provision of suitable earthing arrangements.

Earthing arrangements In days when water authorities used cast iron and copper pipes to provide their service, earthing could be easily effected by means of clamping a current protective

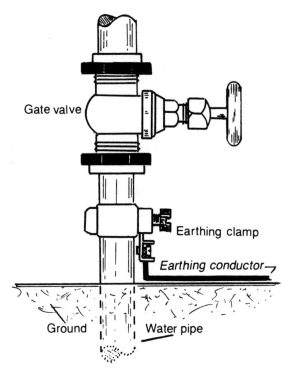

Figure 8.4. An old method of providing an installation earth which can no longer be relied upon.

In areas where soil resistivity is naturally high or the terrain rocky, it becomes less cost-effective to install solid or lattice type copper electrodes in the ground in order to effect suitable earthing arrangements. Because of this, automatic disconnection techniques have been designed in order to overcome this problem. In a modern installation this will take the form of a residual current device (RCD), but in an older or less fashionable system a voltage-operated earth leakage circuit breaker (ELCB) could be found.

Figure 8.5 illustrates a typical TT distribution system in schematic form and highlights the relationship between the current-carrying conductors and the earthing arrangements provided. To recognise this concept more clearly, Figure 8.6 depicts, in graphical arrangement, the correlation between the supply cables and provisions made for earthing for both supplier and customer.

conductor to the supply side of the consumer's stopcock, as illustrated in Figure 8.4. This method can no longer be relied upon, as many metal pipes have now been replaced with modern counterparts made from plastic or polyethylene.

TN-S systems

TN-S supply systems are usually found in urban areas where the distribution of power is dispensed by means of lead sheathed or stranded wire armoured cables laid directly in the ground. Earthing is carried out by simply connecting a suitably sized current protective conductor from the consumer's earthing terminal to the sheath of

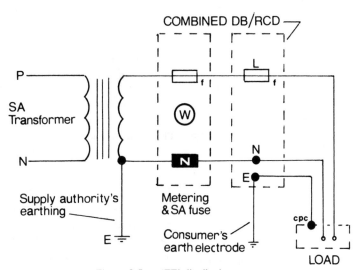

Figure 8.5. A 'TT' distribution system.

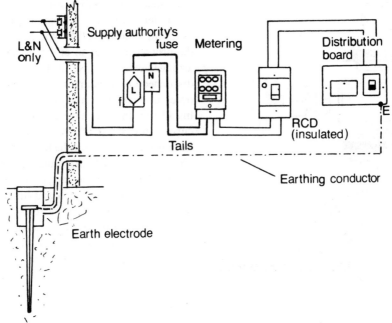

Figure 8.6. A typical 'TT' system shown in diagrammatical form.

the electricity authority's armoured supply cable. This provides a direct route to the star point of the local distribution transformer where it is earthed and ensures that the consumer's fuses or circuit breakers will always operate under fault conditions.

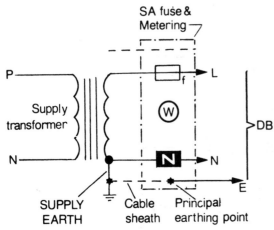

Figure 8.7. A 'TN-S' distribution system. The incoming supply cable sheath provides for a separated earth (see also Figure 8.3).

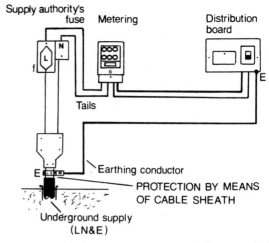

Figure 8.8. A typical 'TN-S' system shown in diagrammatical form.

Earthing arrangements Figures 8.7 and 8.8 illustrate both schematically and graphically the relationship between the supply authority's current-carrying terminations and general earthing arrangements built into a conventional TN-S system. For additional protection, a residual current device may be installed. This will be connected between the electricity supply company's metering equipment and the consumer's main isolating switch. Regulation 413-02-06 refers.

TN-CS systems

The use of a combined neutral and protective conductor to serve a consumer's installation is known as a TN-C system. Should the system be connected to a consumer's installation comprising an independent current protective cable and neutral conductor, the combination is then known as a TN-CS system. This system may be found in both rural and urban areas, distributed both overhead by means of a network of poles, or laid directly in the ground.

Earthing arrangements The schematic arrangement of a TN-C system is shown in Figure 8.9. In practice, the combined neutral and current protective cable is earthed many times at selective points throughout the supply authority's distribution system. A fault condition to earth will, in reality, be a fault occurring between the phase and neutral conductors. This produces a low impedance return path to the star point of the supply transformer, enabling overcurrent protection to operate satisfactorily under adverse conditions.

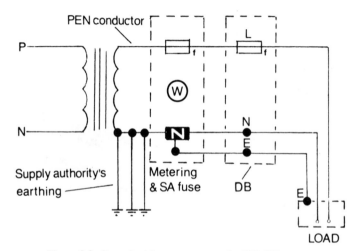

Figure 8.9. General wiring arrangements for 'TN-CS' system.

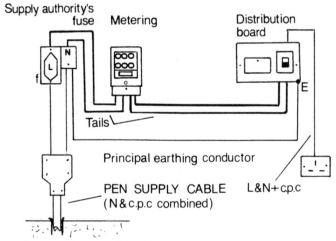

Figure 8.10. A typical 'TN-CS' system shown in diagrammatical form.

Figure 8.10 details in graphical form the earthing arrangements in relation to the supply authority's equipment and the consumer's main isolating switch and distribution system.

TN-C systems

When a supply authority employs a combined neutral and protective conductor in the form of a PEN cable to serve an installation, the arrangement is referred to as a TN-C system.

The consumer's installation will originate from a privately owned isolating transformer then wired throughout using earth concentric wiring. In practice, installations such as these are carried out with single or multicored mineral insulated cable. A special gland pot fitted with a flexible copper lead serves as a combined earth and neutral conductor (Figure 8.11). Figure 8.12 illustrates a typical pot showing the copper lead embedded firmly within its base.

Since the source of the supply is from an isolating transformer there can be no physical connection between the installation and the public distribution system. As an alternative, a privately owned generating plant may be used instead. This method of installation is not in general use but examples of the system can be found serving many types of installation throughout the country.

Earthing arrangements Figure 8.13 shows

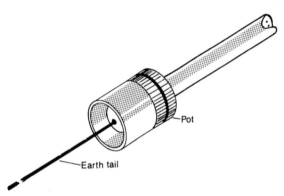

Figure 8.11. Special gland pots, fitted with an integral flexible lead, are used to form a PEN conductor throughout a 'TN-C' installation.

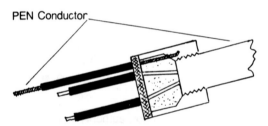

Figure 8.12. Special mineral insulated cable gland pots fitted with flexible copper fly leads are used throughout a 'TN-C' installation.

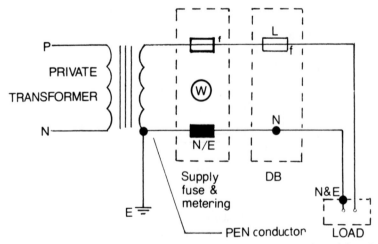

Figure 8.13. General wiring arrangements for 'TN-C' system. Earth concentric wiring is used throughout the consumer's installation.

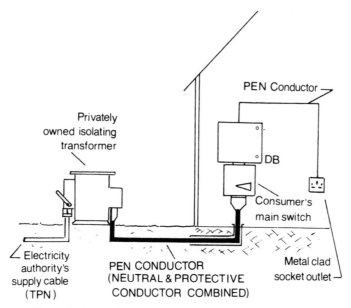

Figure 8.14. A typical 'TN-C' system shown in diagrammatical form.

schematically the relationship between the source of the supply, sub-circuit overcurrent protection equipment and the connected load. For obvious reasons a residual current device cannot be used with this type of installation as both neutral and protective conductor are combined as one. Regulation 413-02-07 confirms.

Figure 8.14 illustrates details of a typical TN-C system shown in graphical form. Combined neutral and earth cables (PEN cables) are used throughout the installation in the form of mineral insulated copper cables.

Handy hints
1. Tools will last longer if they are cleaned and sharpened regularly. Blunt screwdrivers are not only inefficient but can be potentially dangerous.
2. Never leave plant or equipment resting on the rungs of a pair of steps; sooner or later they will come crashing down.
3. When an installation is complete, enter all circuit details on the destination chart serving the distribution centre. This will be a help to others who have to follow you.

Electrical bonding techniques

Bonding must be regarded as one of the most important safety factors in any electrical installation. Without it, lethal currents could cause death or severe injury to an unfortunate victim. There are many environmental conditions where bonding is extremely important but it is often naively overlooked or inadequately carried out. Bathrooms, kitchens, milking parlours, dairies and petrol filling stations are a few examples of areas which cannot and must not be overlooked.

Contact with water will lower body resistance and produce lethal currents even when a lower voltage is present. An average person's body resistance is approximately 6500 Ω. This means that some will have a much higher resistance, whilst others will have far lower. An individual who is saturated by water can reduce this figure by 75 per cent. A child who sustains an electric shock whilst using a humid bathroom could produce a cocktail of lethal currents throughout his or her body.

A combination of stray electrical currents and water can be a deadly partnership. Therefore all areas of high risk, where water or condensation could be present, must be carefully bonded throughout. This applies irrespective of whether an area at issue is served with an electrical installation or not.

Bonding minimises the potential difference which, under fault conditions, could appear between conductive parts and ensures that all extraneous metal-work is electrically placed at zero volts.

Bonding methods and procedures

Two types of bonding are used in electrical installation work and these may be summarised as:

1. Main equipotential bonding (Regulations 413-02 and 471-08).
2. Supplementary bonding (Regulation 547-03).

Main equipotential bonding

This requires all extraneous conductive parts (referring to metal-work not forming part of the electrical installation) and all exposed conductive parts (for example, pressed steel enclosures serving a connected load) to be bonded together in order to place the potential difference of all accessible metal-work at zero volts. This will also incorporate any remotely sited buildings served by the main electrical installation.

Main equipotential bonding conductors will connect all extraneous and exposed conductive parts to the principal earthing terminal, as Figure 8.15 illustrates. This is often furnished as a custom-built integral component but, alternatively, use can be made of the main earthing terminal bar within the distribution centre.

The term 'extraneous conductive parts' can include the following (reference is made to Regulation 413-02-02):

1. Main gas service.
2. Main water pipe.
3. Compressed air supply.
4. Oil pipes serving and forming part of a central heating system.
5. Central heating pipes.

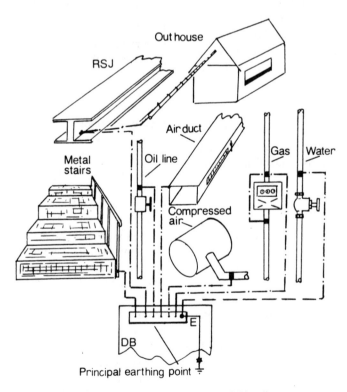

Figure 8.15. An example of equipotential bonding.

6. Exposed steel joists.
7. Air conditioning ducting.
8. Exposed metal infrastructure.
9. Metal stair cases.
10. Metal hand or guard-rails.
11. Steel infrastructure forming a dairy or milking parlour.
12. Metal cladding forming doors or walls.
13. Lightning protective system

Separate current protective bonding conductors may be wired to each extraneous conductive part; the size of the cable being not less than 50 per cent of the main earthing conductor or a minimum of 6 mm in cross-sectional area. Alternatively, one equipotential bonding conductor can be looped from one extraneous conductive part to another, providing the cable is continuous and not broken whenever a termination is formed. In-line joints should be avoided. A continuous conductor will always provide reliable continuity with minimum impedance. The problem of providing a single equipotential bonding conductor to serve an installation can be resolved by slicing approximately 20 mm of insulation from the cable wherever bonding is required. The sliced section of the bare copper conductor can then be folded in

half and placed within a suitably sized pre-insulated compression termination as depicted in Figure 8.16. Reference is made to Regulation 547-02-01.

The *Wiring Regulations* require all bonding terminations to be accessible for inspection, so a little thought has to be given during the 'first-fix' stage of the installation as to where they will be sited (Regulation 526-04-01).

Agricultural situations By their very nature, agricultural installations are open to considerable abuse often inadvertently caused by heavy plant or by the antics of livestock. It is very important that the conductive infrastructure forming a milking parlour, dairy or an all-metal building is effectively bonded to the main earthing terminal. Whichever method is chosen, the equipotential bonding conductor must at all times be mechanically protected from hazardous conditions or from accidental damage.

Cattle are extremely sensitive to electric shock and if subjected to voltages even as low as 50 V can, under defined conditions, die. Always ensure that both bonding cable and clamp are well out of reach of livestock. This will prevent the animals from nibbling the conductors and generally disturbing the bonding clamps.

Gas and water services Main gas and water services must be bonded within 600 mm from the respective supply authority's meter on the consumer's side of the installation. It is wise to shunt both gas and water meters using a suitably sized current protective conductor in order to avoid possible damage should a fault condition occur within a section of pipe adjacent to the meters. Figure 8.17 illustrates this point more clearly. If a water meter has not been included within the installation, the main equipotential conductor must be mechanically clamped as near to the stopcock as practical on the consumer's side of the system.

Check with the local gas and water authorities should special requirements be needed before bonding to their equipment. Regulation 547-02-02 confirms.

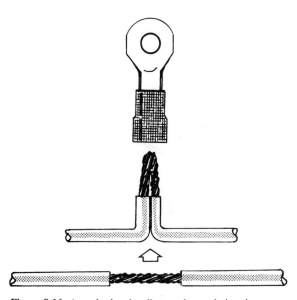

Figure 8.16. An unbroken bonding conductor designed to serve various exposed and extraneous conductive parts.

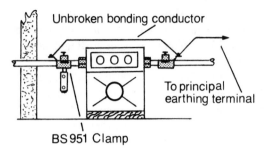

Figure 8.17. Equipotential bonding: shunting gas and, if applicable, water meters.

Special requirements for bathrooms Bathrooms or rooms containing a fixed shower must be effectively bonded to reduce the risk of electric shock. In practice this requires installing a supplementary bonding conductor from the principal earthing terminal to both hot and cold water pipes serving bath or shower and to all extraneous and exposed conductive parts within the bath- or shower-room. Alternatively, a supplementary bonding conductor may be employed to link all exposed and extraneous

conductive parts. This will maintain a common electrical potential of zero volts and greatly reduce the risk of electric shock. Figure 8.18 illustrates this point more clearly.

As an additional safety factor, a bathroom or shower unit can be served by means of a high sensitivity residual current device. In order to accommodate the various circuits a small distribution centre would need to attend the RCD, as Figure 8.19 explains. Regulation 471-08-01 (i) confirms.

Supplementary bonding

This requires all exposed and extraneous conductive parts to be made common with each other as Figure 8.20 illustrates. To meet the demands of the *Wiring Regulations* supplementary bonding conductors are required to have a minimum cross-sectional area of 2.5 mm when mechanically protected or if placed in an inaccessible position, and 4 mm in cross-section if unprotected. In order to be mechanically protected, each conductor will either have to be buried within the plaster line, or alternatively, accommodated

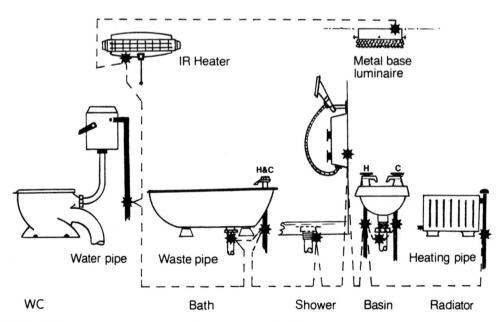

Figure 8.18. Bathrooms and rooms containing a fixed shower must be effectively bonded in order to reduce the risk of electric shock.

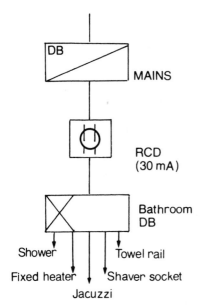

Figure 8.19. A high sensitivity residual-current device may be used to serve a bathroom installation.

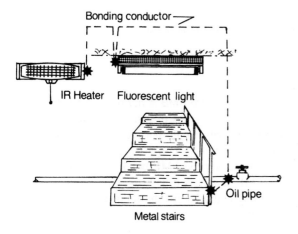

Figure 8.20. An example of supplementary bonding.

within conduit, mini-trunking or capping. Reference is made to Regulation 547-03.

Metal sinks, basins and baths, and hot and cold water pipes together with all extraneous and exposed conductive parts must be made common with each other in order to maintain a uniformly common potential zone.

Radiators Radiators forming part of a central

heating system are generally bonded. If on inspection the plumbing installation is effectively joined by the use of a permanently soldered joint registering low impedance, supplementary bonding need not be carried out. A simple continuity test will establish whether bonding is required, but good practice dictates that it should be carried out irrespective of satisfactory continuity readings. After all, it is better to 'err' on the side of caution where safety is concerned — though others may disagree! Reference is made to Regulation 130-04-01.

Although supplementary bonding can, under certain defined conditions, be overlooked, a central heating system will still require to be equipotentially bonded.

Special bonding requirements for protective multiple earthing (PME) supplies

The use of a combined neutral and earth distribution system requires special approval from the Department of Energy. This not only imposes legal requirements on a supply authority but on a consumer as well. For a consumer, this requires special bonding arrangements to be carried out within the installation. The method adopted to form an equipotential zone has been debated in previous paragraphs.

Principal PME earthing conductors
The size of the main earthing conductor forming part of a PME system is directly related to the cross-sectional area of the incoming supply conductor. Main equipotential bonding from a consumer's distribution centre to principal extraneous conductive parts is increased in size and again is correlated to the supply authority's earthing conductor.

In practice, if the size of the supply authority's main neutral conductor serving the metering equipment is either 16, 25 or 35 mm in cross-sectional area then the consumer's principal neutral conductor catering for the distribution centre will be the same. The size of the main equipotential bonding conductor is increased to 10 mm in cross-sectional area. Figure 8.21 illustrates this more clearly. Should the size of the main neutral

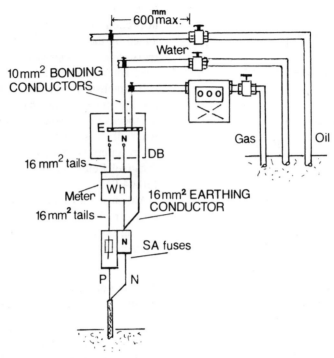

Figure 8.21. An example of protective multiple earthing requirements.

conductor serving the customer's switchgear be 50 mm in cross-sectional area (csa) then the size of the earthing conductor for final connection into the supply authority's system from the consumer's distribution centre need only be 25 mm in csa. The size of the bonding conductor between extraneous conductive parts and the supply authority's PME terminal is then increased to 16 mm in csa. Reference is made to Table 54H.

If at all in doubt, guidance may be gained from any local supply authority's administrative offices. It is far better to seek advice than to get it wrong.

Additional protection A residual current device may be used for additional protection if required or considered necessary. Fault voltage circuit breakers (FVCB) must never be used in conjunction with a PME installation. The low neutral/earth impedance path developed by this system will render the device inoperative by effectively short circuiting the operating coil.

Installations where PME is not recommended
There are many types of installations where a supply authority would not recommend a protective multiple earthing sysem, arising from difficulty in trying to satisfy the PME bonding requirements.

A PME supply is usually refused by an electricity authority for the following types of installation:

1. Caravans and mobile homes. When a building has been fabricated using a metal frame or structure, a PME earthing terminal will not be provided. A residual current device is recommended.
2. Milking parlours and dairies, etc. Difficult to satisfy PME bonding requirements. Residual current device recommended.
3. Mines and quarries. Special earth leakage measures are required to comply with the *Mines and Quarries Electrical Regulations, (1956).*

4. Building sites. Very difficult to satisfy PME bonding requirements. A high sensitivity residual current device is recommended.
5. Petrol filling stations. Many practical difficulties in satisying the PME bonding requirements. Must be effectively bonded, including all pipes, tanks and man-hole covers, etc. An RCD is recommended.
6. Swimming pool. The installation as a whole may be served with a PME earth terminal but the supply to the swimming pool should be separately protected with a suitable residual current device.

Circuit protective conductors

Circuit protective conductors (cpc) may appear in many different forms, depending on the design of the installation they are serving. Solid drawn steel conduit provides a very suitable protective conductor as long as all joints are secure and installed correctly. The nominal cross-sectional area of 25 mm diameter is approximately 131 mm^2, whereas 32 mm diameter conduit yields a cross-sectional area of 170 mm^2. This is, in the majority of installations, more than adequate and will provide an excellent current protective conductor. Steel cable-trunking systems can offer an even greater advantage. Installations constructed

using 150 × 150 mm diameter metal trunking will provide a current protective conductor of approximately 750 mm^2. Care should be taken throughout construction to maintain both mechanical and electrical continuity. Each section of trunking must be firmly assembled and bridged with a suitable copper earth link as illustrated in Figure 8.22. It only requires one badly made joint to produce a high level of impedance and this, coupled with a fault condition to earth, will create a very dangerous and unreliable installation.

Choice of protective conductors
There are many different types of circuit protective conductors available but choice is usually limited to the design of the installation. Listed below are a selection which are in general use. Regulation 543-02-02 confirms.

1. The steel wire protecting armoured cables.
2. The copper sheath of mineral insulated cables.
3. The covering of lead sheathed cables.
4. Steel cable trunking, providing all joints are mechanically sound.
5. Heavy gauge welded steel conduit, other than flexible steel conduit, where an independent current protective conductor must be provided. Regulation 543-02-01 confirms.
6. Single core non-sheathed green/yellow general purpose cable. Uninsulated conductors should not be used.
7. The integral current protective conductor within PVC insulated and sheathed cable. Connections must be mechanically sound and of low impedance.
8. Copper tape.
9. Continuously formed extraneous conductive parts.

Disconnection times: protection against electric shock

In order to offer a realistic degree of protection against electric shock, Regulation 413-02-09 and Table 41A require the earth loop impedance (Figure 8.23) to be sufficiently low at every socket outlet, to allow for automatic disconnection to

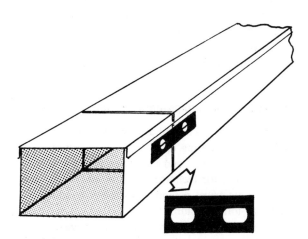

Figure 8.22. Each section of steel cable trunking must be bridged with a copper earth link and secured by means of shakeproof fixing screws.

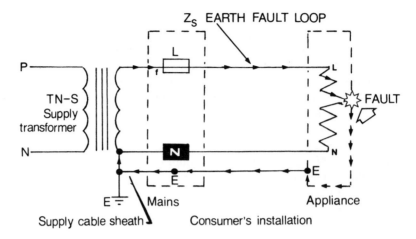

Figure 8.23. The total value of Z_s, the earth fault loop impedance, will vary depending on the distance the installation is from the supply transformer or, in cases where a supply earth is not provided, the type and consistency of the sub-soil.

occur within 0.4 seconds. Should the installation be serving an agricultural, horticultural or temporary premises, the disconnection time is reduced to 0.2 seconds. Disconnection times are satisfied when the nominal voltage is between 220 and 277 V and is served from a TN supply system. This can be tested by use of an analogue or digital *loop impedance meter*. Should the test prove unacceptably high, first check the cpc connection serving the accessory under review. If satisfactory, effective supplementary bonding will help reduce the value to an acceptable level. This type of situation is relevant when the area served by the installation is large and the wiring is carried out using PVC insulated and sheathed cables.

Metal-clad, hand-held electrical appliances are potentially dangerous under earth fault conditions so it is imperative that the supply is automatically disconnected swiftly in the recommended time. Generally, for fixed equipment such as lights, central heating controls, fixed wall storage heaters, fans, electric motors, etc. the disconnection time must not exceed 5 seconds.

Automatic disconnection
Automatic disconnection due to a fault condition to earth can be afforded by means of:

1. Overcurrent protection

   (a) miniature circuit breakers

   (b) moulded case circuit breakers
   (c) fuses
   (d) overload relays

2. Earth leakage protection

   (a) residual current device (RCD)
   (b) fault voltage circuit breaker (FVCB)

Whenever a residual current device is employed as a principal means of disconnecting the supply, it should be kept in mind that high sensitivity types are far more preferable as a means of protecting circuits incorporating socket outlets. Residual current devices respond to a leakage of current from either phase or neutral via conductive parts to earth and therefore are totally unsuited as a means of protection against overcurrent.

Earth-free local equipotential bonding

This is a very specialised arrangement of equipotential bonding often found in research centres, laboratories and industries where it is essential that, for technical reasons, operatives are the same electrical potential as the surroundings, plant and infrastructure.

This method effectively bonds all extraneous conductive parts together, but not to earth. This is called earth-free local equipotential bonding. In view of its specialised nature, this technique may only be carried out when specified by an electrical engineer and only when certain conditions are met. For the majority of installations, earth-free equipotential bonding would be totally unsuitable and would only be carried out in exceptional circumstances and only when authorised by an electrical engineer. Reference is made to Regulations 413-05 and 471-11-01.

Summary

1. There are four principal methods of distribution in Britain; all of which must comply with the *1937 Electricity Supply Regulations*. This, in general terms, requires the source of the supply to be earthed at one point within the network. TT, TN-S, TN-CS and TN-C systems all meet this requirement.
2. TT systems are usually found in rural areas, distributed by means of a network of overhead cables. The system requires the consumer to provide an effective means of earth leakage protection. This may take the form of a residual current device or alternatively a lattice type earthing arrangement buried in the ground.
3. TN-S systems are distributed in urban areas. The sheath of the underground service cable, acting as a current protective conductor, is connected to the star point of the distribution transformer and earthed.
4. A TN-CS system provides a combined neutral and earth cable (PEN cable) to serve an installation having both separate and independent current protective and neutral conductors. This type of installation can be found in both urban and rural areas.
5. TN-C systems are generally in decline and are only used in conjunction with a privately operated isolating transformer or generating plant. All installation work is carried out using cable having a combined neutral and current protective conductor. This usually takes the form of minerally insulated cables.
6. Bonding minimises the potential difference appearing between extraneous and exposed conductive parts. Once effectively bonded, all metal infrastructure forming part of, or extraneous to, the electrical installation is placed at a common potential of zero volts.
7. There are three accepted methods of bonding:

   (a) main equipotential bonding;
   (b) supplementary bonding;
   (c) earth-free local equipotential bonding.

8. Main equipotential bonding connects all extraneous conductive metal-work, such as gas and water supplies, to a principal earthing point.
9. Supplementary bonding requires all exposed conductive parts to be made common with all extraneous conductive parts. In practice, this requires baths, basins and fixed electrical appliances to be bonded together as Figure 8.24 illustrates.
10. A high sensitivity residual current device may be used for additional protection to serve hazardous areas.
11. The use of a combined neutral and earth supply, often called a protective multiple earthing system, imposes a mandatory requirement on the consumer to bring the bonding arrangements to a satisfactory agreed standard. This requirement can vary throughout the country. A check with the local supply authority would be prudent before commencing work.
12. PME supplies are not recommended to serve:

   (a) milking parlours;
   (b) farms;
   (c) mines and quarries;
   (d) building sites;
   (e) swimming pools;

   or where difficulty may be experienced satisfying PME bonding requirements. Instead, a suitably sensitive residual current device should be used.
13. Depending on the nature of the installation, current protective conductors will take many different forms. These include:

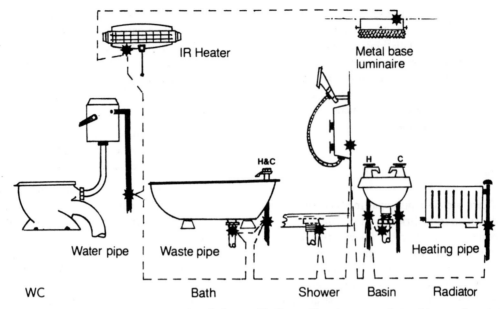

Figure 8.24. Supplementary bonding requirements for a bathroom. Ideally, earthing clamps manufactured in accordance to BS 951 should be used as a method of making a bonding connection to the pipework.

(a) the outer sheath of metal-clad cables;
(b) steel conduit;
(c) steel trunking systems;
(d) copper tape;
(e) single core, non-sheathed, green/yellow PVC insulated copper cable;
(f) continuously formed extraneous conductive parts.

14. In the event of a fault condition to earth, automatic disconnection time must occur within:

    (a) 0.4 seconds for installations serving socket outlets or hand-held electrical appliances.

    (b) 5 seconds for fixed electrical equipment. This may be afforded by means of circuit breakers, fuses or residual current devices.

15. Earth-free equipotential bonding may be carried out only when specified by an electrical engineer. This technique effectively bonds all extraneous conductive parts together; but not to earth.

Handy hints

4. It is often easier to attach bonding conductors to earthing clamps during the first fix period of an installation when site conditions are less cramped.

5. Never use aluminium conductors to make final terminations to copper earthing rods or plates. Corrosion will set in and contaminate the mechanical bond.

6. Socket outlets, other than shaver points, must not be installed in a room containing a fixed shower or bath. Where a room contains a fixed shower cubicle, socket outlets must be at least 2.5 m from the cubicle.

7. The service side of a water main should never be relied upon as an effective earth. Much of the original plumbing has been replaced with plastic or polythene counterparts.

• The term SELV was originally formed using the initial letters of the electrical definition, Safety Extra Low Voltage. It was then decided that reference should not be made to the meaning of the individual letters as no voltage should be regarded as safe. However, the letters are now defined as Separated Extra Low Voltage; that is, circuits which are separated from earth and from other systems.

- The nominal domestic and industrial voltages were changed from 415/240 volts, ± 4% to 400/230 volts, +10%, −6% on 1 January 1995. By the year 2003 UK voltage tolerances will be changed to ±10%. The wider tolerance levels permitted will mean that domestic voltages will be between 207 and 263 volts.
- The *thermistor*, an electronic semi-conductor component made from a compound of oxides, manganese and nickel, will rapidly decrease in resistance from about 0.1 MΩ at

293 Kelvin (20 °C) to a mere 10 Ω at 373 Kelvin.

It is used to compensate for temperature variations which could cause an increase in resistance to other components in a sensitive electronic circuit. The thermistor is often used as a sensitive thermometer.

- The *thyristor*, or *silicon controlled rectifier* is a semi-conductor component used in circuits for the control of mercury arc rectifiers. Conduction will only take place when a third terminal, called a *gate* or *base*, is positively fired in respect to the terminal attached to the outside '*n*' layer. Once conduction has taken place current flow will be maintained until such time that the applied voltage is reversed or falls to zero.

9 Practical lighting and power circuits

In this chapter: Alternative lighting circuitry, advantages and disadvantages. Power circuit installation techniques. Floor traps. Kitchen installations. Secrets of second fix.

Polyvinyl chloride insulated cables

Polyvinyl chloride (PVC) insulated and sheathed cables were first introduced to industry after the Second World War. Conductor sizes range from 1.0 to 400 mm² in cross-sectional area and are able to operate safely in temperatures up to 343 Kelvin, 70 degrees Celsius.

PVC insulated and sheathed cables are made to suit installation needs by offering a choice of conductor cores; the most common being twin and three core. All cables carry a bare current protective conductor.

The use of this type of cable offers a wide choice of wiring methods especially when installing lighting and power installations. Each method has its advantages and disadvantages and it is these which will be evaluated next. Also comparisons will be made between these methods and traditional industrial methods of wiring. Figure 9.1 outlines the basic requirements for a typical one-way lighting circuit.

The basic lighting circuit

A suitably sized lighting grade cable, 1 or 1.5 mm², with matching overcurrent protection is wired from a local distribution point, terminating at a ceiling outlet.

A modern ceiling rose is constructed, forming three independent sections where conductors may be safely terminated. The middle section is reserved for the incoming and outgoing phase conductors, whilst the neutral is connected and ranged to the left. At this point the neutral conductor is adjacent and common with the neutral flexible cable serving the lampholder as Figure 9.2 illustrates.

Wiring to the switch
A separate cable is then wired from the ceiling outlet to serve the control switch where both conductors and the current protective cable are terminated. The black switch wire must be phase colour-coded at the ceiling rose and switch in order that its status may be altered and to provide

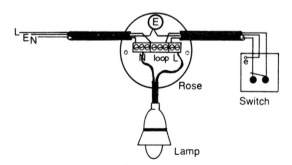

Figure 9.1. The loop in method of wiring a one-way lighting circuit.

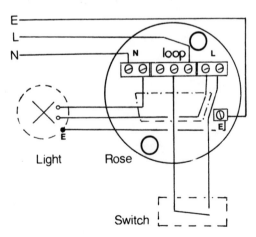

Figure 9.2. The supply neutral is common with the neutral serving the lampholder.

means of identification. This may be achieved by use of either coloured PVC adhesive tape or plastic sleeving. Reference is made to Regulation Table 51A.

At the ceiling rose, the red switch feed conductor is connected and made common with the incoming phase conductor ranged in the middle section. The black, colour-coded, switch wire is then ranged to the right of the centre and terminated adjacent and made common with the other flexible cable serving the lampholder.

All protective cables are sleeved using PVC green/yellow oversleeving and made common with each other within the termination provided. Regulations 514-03-01 and 514-06-02 refer.

Lighting installation methods

Seven different methods of wiring which could be incorporated into a lighting installation are now discussed. Some are commonplace and are carried out every day, whilst others which may appear unfamiliar are only employed under certain defined conditions:

1. Joint box method.
2. Loop in method.
3. Central joint box system.
4. Double pole switch method.
5. Metal or PVC conduit system.
6. Final ring circuit method lighting.
7. Fused connection unit method.

Joint box method
Figure 9.3 demonstrates how a lighting circuit may be wired by use of the joint box method (reference is made to Regulation 130-02-05). A single phase supply of suitable size is taken from a local distribution board to serve a single or series of joint boxes within an area chosen for the circuit. Cables entering or leaving the boxes, providing means of control or delivery, are terminated internally as illustrated in Figure 9.3.

The advantage of this system is that the majority of cables may be laid and connected during the first fix stage of the installation. It is usually agreed to be quicker and easier to connect electrical accessories as there are only one or two cables ever present, the majority being connected

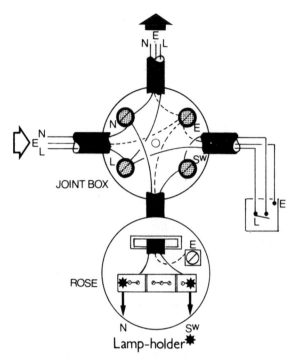

Figure 9.3. Lighting installation: joint box method.

into the joint boxes during the construction or first fix stage.

The disadvantage of this method usually materialises after the installation has been commissioned. It can be very difficult to trace and identify faults or to carry out a progressive cable test when, for example, a chipboard floor has been laid or a decorative interlocking flooring installed. Another weakness is the practical accessibility of all the joint boxes. There is a need to know where they lie in order to test or add to the original installation. Ideally, marked traps cut and fitted directly over the joint box during the construction stage would provide a solution for future maintenance and installation work. Sadly, the electrical installation is usually completed long before any practical arrangements may be made for cutting traps. Figure 9.4 illustrates how a small inspection trap can be constructed in traditional flooring in order to gain access to an electrical joint box.

Industrially Industrial techniques could be adapted to carry out the same installation but it would

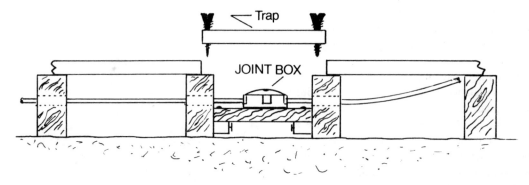

Figure 9.4. Providing access to a joint box by means of a floor trap.

prove very costly and time consuming. If conduit were to be used, any future rewire could be undertaken with the minimum of distruption as all outlets could be designed to be completely accessible from one level. The need to remove flooring could be dispensed with.

The loop in method

The loop in, or three plate ceiling outlet, method is one of the most widely used techiques in electrical installation work today.

Using PVC insulated and sheathed cable, a suitably sized conductor is taken from the main distribution centre and routed to provide a direct feed to each lighting point. The circuit is run in a manner similar to the joint box method discussed in previous paragraphs; but there the similarity ends. All wiring connections are made at either the lighting point or the control switch as shown in Figure 9.5.

There can be as many advantages as disadvantages using this method of wiring. The main benefits are ease of wiring and that all terminations are completely accessible from one level. Fault-finding and testing can be undertaken without the need to remove flooring so technical problems can be quickly traced. The installation may be added to without possessing prior knowledge as to where the joint boxes are situated, unlike some other procedures. The advantages gained by employing this method can often outweigh other techniques. An installation fault developing in a newly carpeted dwelling could prove somewhat embarrassing if, for example, the joint box method had been adopted.

Unfortunately, with this method cables can be

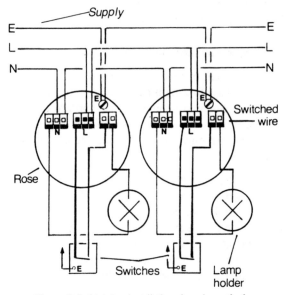

Figure 9.5. Lighting installation: loop-in method.

accidentally pulled out of the ceiling during the first fix period of the installation and cause a great deal of trouble when accessories are fitted. Should there be any danger of this happening, tie or twist the cables together so they provide support for each other. Wiring ceiling points can take up to three times longer and mistakes can be made if the switch wire (the cable from the switch to the lighting point) is incorrectly marked or recognised. Always identify cables clearly. Prepare the switch wire first and attach either a red sleeve or adhesive PVC tape around the black switch wire as Figure 9.6 illustrates. This will help to ensure that the lighting point is wired correctly.

It is very tempting to overload one particular

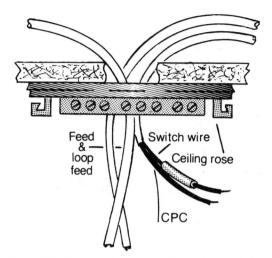

Figure 9.6. Prepare and colour-code the 'switch wire' first.

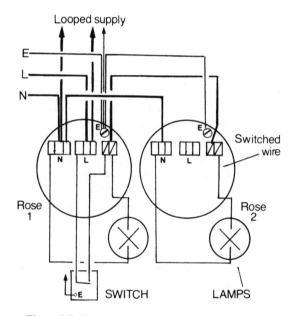

Figure 9.7. Two lighting pendants controlled by a single switch.

when wiring is carried out for wall lights. Too many will create a practical disadvantage should the fitting be slim or small.

Outdoor fittings Outdoor and storm-soaked electrical fittings wired using the loop in method can often cause tripping problems when the installation is served by a residual current device. Under these circumstances it is far more sensible to fit a double pole switch to serve the outside lighting installation as both phase and neutral conductors are disconnected from the circuit whenever the switch is placed in the 'off' position. Figure 9.8 illustrates this in schematic form.

Central joint box system

Many electricians prefer the central joint box system of wiring to other well-established methods. Work on the majority of cable terminations may be carried out in one operation and in one place, choosing whether to connect during the first or second fix stages of the installation. Figure 9.9 outlines this method of wiring.

An appropriately sized lighting cable is fed from a lighting distribution board to serve the central joint box. This usually takes the form of an adaptable box made from plastic and measuring approximately 100 × 100 mm by 25 mm in depth. The supply cable is terminated into a suitably sized connector block which is secured to the base of the box. Additional circuits, fed from the same phase, may be added to the system; each circuit occupying an allotted set of connectors.

The advantage of this arrangement is that the majority of interconnections may be carried out at one or two points within the installation at a time convenient to the operative. This method is very simple, providing all the cables are identified.

lighting point with supply and looped supply cables when positioned in a suitable or handy place. This practice will cause cable crowding. Restrict yourself to one cable serving the outlet, a looped feed to the next light and a control cable to the switch. Sometimes there may be more if, for example, two or more lights are controlled by the same switch. Figure 9.7 illustrates this point more clearly. Cable crowding can inadvertently occur

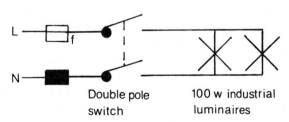

Figure 9.8. A double pole switch can be installed to serve an exterior installation.

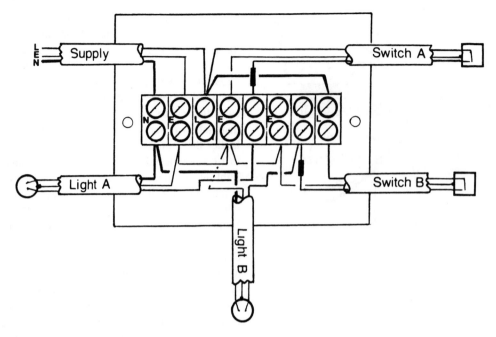

Figure 9.9. Lighting installation: central joint box method.

Should the cables be installed unmarked, much time is wasted in re-establishing the role of the cables. As only one cable is used to serve either lighting or switch point from the central joint box, cable crowding, as experienced with the loop in method, is lessened.

A floor trap which is both accessible and easily identified should be constructed to serve the central joint box. This will enable maintenance and additional work to be carried out. Installation faults and problems often occur as a direct result of wrongly marked cables or cables which have not been labelled at all. The golden rule must be: *always label your cables*!

Industrial use It is unusual to carry out this wiring method when cables are drawn through conduit and trunking installed to serve industrial installations. Should the installation be wired using minerally insulated cable, the central joint box system would then be appropriate. The plastic joint box used to serve the PVC insulated and sheathed cables would be exchanged for a suitably sized pressed steel box in order to provide earthing continuity throughout the installation (see also Chapter 11).

Double pole switch method

With this type of lighting installation the double pole switch is employed in two roles: firstly, as a means of control and, secondly, to connect and terminate incoming and outgoing supply cables throughout the circuit. Figure 9.10 illustrates the basic requirements for such a method.

This system is often employed for wiring portable accommodation where it may be impractical to use joint boxes or the loop in method of carrying out an installation. As with other techniques the double pole switch method has many advantages and disadvantages. It is very easy to test, so fault- and problem-finding is both simpler and quicker. As with the loop in system all cable terminations are on one level and with the absence of joint boxes, mistakes in wiring can be minimised.

Cable crowding may be completely avoided as only one or two cables will be present at a lighting point or two or three at a switch position.

The double pole switch method is ideal to serve an outside lighting installation which is monitored by a residual current device. Both phase and neutral conductors are disconnected from the supply when the switch is placed in the 'off'

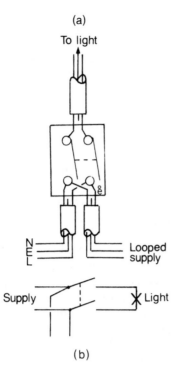

(a)

To light

N
E
L

Looped
supply

Supply — Light

(b)

Figure 9.10. Lighting installation: double pole switch method (a), shown in schematic detail (b).

position, enabling any fault condition to be temporarily removed from the circuit. Wiring by this method is simple to install, but tends to be more expensive compared to other wiring systems as more cable is used and additional funding is required for the use of double pole switches in the installation. It is obviously impossible to wire a two-way and intermediate lighting circuit by the use of double pole switches, so this method of control will have to be carried out in a more traditional manner. Switch boxes need to be deeper to accommodate the extra cables and it is often more difficult to add additional circuitry once the installation has been commissioned.

Industrial use The double pole switch method is seldom used in industrial lighting installations but may be found in special unconventional installations calling for the mechanical isolation of both the phase and neutral conductors.

Steel or PVC conduit method
Industrial, commercial and agricultural lighting

installations are usually carried out using solid drawn steel conduit or PVC-based conduit. Whatever the choice, conduit systems have an unquestionable advantage over most other wiring methods because of their adaptability. Circuits may be added or removed with the minimum of trouble. Extending the installation may be achieved without too much difficulty, though care must be taken to ensure that the original cables are not harmed. Damage usually takes the form of *cable burn* and can develop when PVC insulated cables are pulled through conduits which are close to their maximum cable capacity. To reduce the effect of friction, apply powdered french chalk to the cables to be installed before drawing into the conduit.

Solid drawn conduit is both mechanically strong and robust and is to be found throughout industry. The most widely used sizes are 20,25 and 32 mm in outside diameter. Larger sizes are available and may be found in heavy industry or serving installations with special needs. Other than specialised requirements, solid drawn metal conduit is manufactured in a choice of either black enamel or galvanised finish. Support is by means of saddles and clips mounted at regular intervals throughout the installation. A selection of screw-on accessories is available and known throughout the industry as BESA (British Electrical Standards Accessory) boxes. Figure 9.11 illustrates a selection of boxes most commonly used today.

Installation techniques will be discussed in Chapter 10.

Wiring techniques: conduit Wiring is carried out using PVC or vulcanised indiarubber (VIR) insulated single cables drawn through conduit. The operation may be aided by use of a nylon or sprung steel draw-in tape. For ease of wiring, the insulation surrounding the cable ends must be stripped to expose approximately 75 mm of copper conductor. These are then attached to the draw tape and to each other in a staggered fashion producing a spearhead of cables. Figure 9.12 helps to clarify this technique. A streamlined spearhead of cables will enable the task to be undertaken far more easily than if the cables are just tied on to the draw wire as illustrated in Figure 9.13. It is

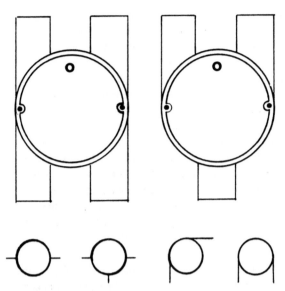

Figure 9.11. British Electrical Standard Accessories (BESA) for conduit installations. From left to right, top row: twin through way; branch three way; bottom row: through way; three way; angle tangent; branch 'U'.

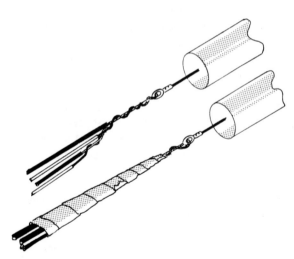

Figure 9.12. Preparing cables to be drawn through conduit. The prepared cables may be lightly taped in order to avoid snagging.

Bunched cables

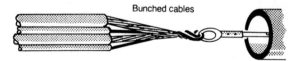

Figure 9.13. The wrong way to prepare cables for drawing through conduit.

most important to ensure that all the cables are firmly and securely attached to each other; free from loose and jagged ends which might hamper their passage through the conduit.

Should there be insufficient inspection boxes throughout the installation, navigating a draw tape through empty conduit could prove troublesome. If sticking occurs and progress is halted, try tapping the conduit with a hammer whilst a colleague pushes the draw tape. This is often quite sufficient to free the lodged tape. However, if unsuccessful try pushing an auxiliary draw tape taking the form of a short length of cable, bared and hooked at the leading end from the opposite direction. By passing the head of the nylon draw tape and twisting the auxiliary cable, the hooked section will attach itself around the main draw tape. This again is a four-handed job, one person pushing the main tape whilst the other gently pulls the hooked auxiliary cable. This technique usually succeeds when other methods fail. Figure 9.14 helps to explain this further.

For ease of wiring always install sufficient inspection boxes throughout the installation and problems involving seized draw tapes will be minimised.

Once carcassing has been completed and the installation is ready for wiring, the phase conductor is taken from a local distribution board to serve the control switch whilst the neutral is wired directly to the lighting point. Should there be any design need to continue the lighting supply to another area within the installation (Figure 9.15), the feed may be looped from these two points. To complete the circuit a switch wire is taken from the switch to serve the lighting point. Had the installation been carried out using PVC conduit, a green and yellow striped current protective conductor would have been included in the circuit (Regulation 514-03-01).

This is the basic method adopted when wiring conduit installations, clarified in Figure 9.16.

Figure 9.14. A single conductor, hooked at the leading end, will help to free a snagged nylon draw-in tape.

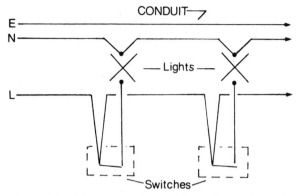

Figure 9.15. Lighting installation carried out using steel conduit (schematically drawn).

Advantages and disadvantages There are many advantages and disadvantages of choosing a PVC or steel conduit installation. The advantages include the following:

1, Rewiring is made easy.
2. Both power and lighting circuits may be accommodated in one conduit.
3. May be added to with ease.
4. Can be used in a hazardous environment when used with special fittings.

5. Robust; will resist mechanical damage.
6. Different sized conduits may be employed within the same installation.
7. May be installed indoors or outdoors.
8. Different coloured cables may be used to suit the needs of the installation.
9. Can be cast in concrete.
10. A good choice of accessories available to suit the needs of the installation.

The disadvantages may be listed as follows:

1. High installation costs.
2. Can cause technical problems should the installation be badly installed with disregard to earth continuity (see Chapter 2). Reference is also made to Regulation 543-02-04.
3. Plastic conduit is not flame-proof.
4. Can be over-accommodated with cables (Regulations 522-08-02 and 06).
5. Black enamel conduit will rust in an unsuitable environment.
6. Plastic conduit requires a current protective cable for each circuit accommodated. A plain green coloured conductor should not be used. Regulation 514-06-02 refers.

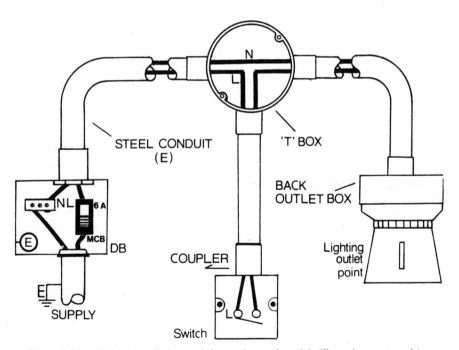

Figure 9.16. A lighting installation carried out using steel conduit (illustration not to scale).

7. Plastic conduit will sag in direct sunlight if installed facing a window or if placed in a warm environment.
8. A high level of skill is needed.

Wiring It is most important that all cables are drawn into the conduit together. This will virtually eliminate cable burn. It often requires considerable planning on behalf of the electrician should the installation be complicated or troublesome. It is wise to sit down and think things through before attempting to wire as it is easy to get things wrong!

Mixed voltages may be included within the same conduit, providing the value of the insulation of the lowest voltage cable matches the insulation value of the highest voltage conductor. As an example, a typical 6 V bell would never be installed within the same conduit as a 230 V lighting circuit (Regulation 528-01-01 confirms).

Generally, wiring requires two operatives: one to pull the draw-in tape whilst the other feeds the cable into the conduit. Each cable drawn in must be physically parallel with the other to reduce the risk of snagging around bends and sets. Both operations must be carried out in complete unison in order to achieve any measure of success.

Never attempt to draw in too much cable in one operation. Form loops of cable at convenient inspection boxes and feed one loop of cables into the conduit at a time. A far greater control may be achieved when the installation is wired in this method. Figure 9.17 will help to make this more clear.

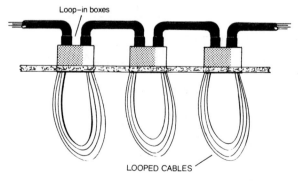

Figure 9.17. Use cable loops when wiring a conduit installation.

Dry lubrication Finely powdered french chalk is an ideal aid when drawing cables through conduit. Should the installation prove stubborn, try dusting a small amount on the surface of the cables or blow french chalk around the bends and sets using suitable size oversleeving. This will help to reduce the resistance between the cables and the internal surface of the conduit and will assist in drawing them in. Never use grease or petroleum jelly.

Switch wiring identification To avoid confusion, especially when making off multigang switches, nick the switch wires leading to the various lighting points throughout the circuit. By keeping the feeds unmarked and coding the switch wires, individual cables may be readily identified when final connections are carried out.

Care should be taken when connecting aluminium conductors into switch terminals. Being softer than copper, aluminium cables are easily compressed by the pressure from the terminal screw when tightened and may shear off unnoticed, only to bring problems after the installation has been commissioned. Never use aluminium conductors in a damp environment; it could cause electrolytic action. Regulation 522-05-02 confirms.

Volt drop in systems employing steel or PVC conduit Be wary of potential problems involving volt drop arising from lengthy cable runs serving inductive lighting loads. Should the drop be unacceptable, then both cables and conduit will progressively increase in temperature causing problems or damage to the installation over a period of time. If in doubt, always wire in a larger sized conductor but, better still, calculate the potential volt drop mathematically. Consideration should be given to correction factors and a suitably sized cable selected from figures obtained.

Final ring circuit method: lighting

This method of wiring is not commonly used in Britain, mainly due to the additional cost of materials required. Figure 9.18 illustrates in schematic form details of how the circuit is wired. This technique may be applied to a variety of installations. In a domestic situation, a final ring

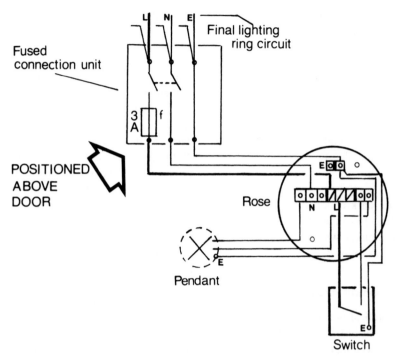

Figure 9.18. Lighting installation: final lighting ring circuit method.

lighting circuit may be wired to serve an area where lights are required. Instead of socket outlets, unswitched fused connection units are used and sited in the centre and above each door. Wiring is then reverted to the loop in or three plate ceiling outlet method to serve both switch and rose as described in 'The loop in method' section. A PVC insulated and sheathed cable is taken from the 'load' side of the fused connection unit to the lighting point. A switch wire is then routed from the ceiling outlet to the switch position and interconnections are made within the ceiling rose. All 13 A cartridge fuses are removed from each fused connection unit and replaced with sizes suited for the load in each room or area served.

Although expensive to install, this technique of wiring has an advantage over most other methods. Should an overcurrent occur, only the room or area served by the fused connection unit will be affected and the circuit may be quickly repaired.

Fused connection unit method

Lighting points serving special needs may be taken from a fused connection unit forming part of a final ring circuit. Where other methods prove costly or impractical, a lighting point wired by this method will provide a practical solution to the problem. Special circumstances will often dictate whether or not to employ this technique of wiring. Take as an example, a remotely situated area previously wired for power in the form of a single 13 A socket outlet serving an extended final ring circuit. Installation of additional cables to supply a lighting point could prove both costly and time consuming.

Figure 9.19 illustrates the basic cabling

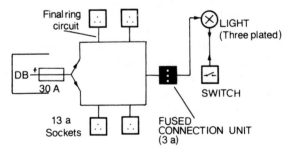

Figure 9.19. Lighting installation: final ring circuit method (power).

requirements when wiring in this manner. A suitably sized cable is taken from the load side of the fused connection unit and directed to the lighting point. The control switch is wired and then routed back to the luminaire where all interconnections are made. This, or any of the other methods previously debated, may be used to wire from the fused connection unit. After completion, protect the circuit with an appropriately sized cartridge fuse.

Handy hint
- Remember to mark or identify your cables during the first fix stage of an installation. You will be familiar with their function, but will others who follow?

Practical domestic power circuits

A selection of theoretical wiring systems have already been examined in Chapter 6 and popular wiring methods and techniques considered.

Inexperience can be the cause of much frustration when unfamiliar work is undertaken. Final ring circuits, although theoretically straightforward, are often accompanied by many practical problems. With this in mind we will be examining practicalities involved when work is carried out involving such an undertaking. A planned installation will be far better managed than one carried out with little thought.

Remember the acronym 'STRAW': Stop, Think, Review And Work. Mistakes cost time, money and very often pride!

Wiring
Always start at the highest point. In a typical three bedroomed house this will probably be the loft or roof space. Often it is quicker and easier to install the first floor final ring circuit from a position within the loft. Cables serving the circuit may be routed from the distribution board under protective capping or drawn through oval conduit to the roof space. Once within this area, cables should be clipped every 600 mm or so to the side of the roofing timbers. Never be tempted to lay cables within the roofing space unclipped. They are easily damaged or disturbed when laid on top or over roofing timbers.

Clipping may be made easier and more interesting by smoothing out the PVC insulated and sheathed cable for a distance of approximately 3 m at a time and fixing the prepared run with an 'anchor clip'. Clipping between the anchored clips is then straightforward and can be made to look neat. When the installation is surface mounted, try smoothing out the cable using the thumb once every metre, spacing the clips approximately every 200 mm. Continue to smooth out the cable each time a clip is secured to obtain a professionally installed look to the installation.

Should the cable have to be installed on the surface of a hard brick wall, it is often difficult to fix the clips without the nails either bending or snapping off. If this should happen, try vibrating the clips into the wall between the edge of the brickwork and the mortar with a medium sized hammer. To avoid damage to the eyes, always wear appropriate eye protection when driving clips into hard masonry or brickwork. The hardened nails can snap off and fly, causing physical harm if not hammered home squarely.

At times it may be necessary to route a cable through a dividing wall. If an electric drill is unavailable, a 20 or 25 mm diameter metal conduit with a 'V' slot cut in one end makes an excellent substitute hole maker. This is ideal for soft walls and masonry but is not recommended for concrete. Once a small niche has been established, keep rotating the conduit until the hole has been made. If the conduit is not turned whilst being hammered it will probably seize within the wall. Figure 9.20 illustrates how such a tool may be constructed. Before the cable is routed through the hole, install a short length of plastic conduit in order to protect the installation.

Regulations 527-02-01 and 03 demand each end to be sealed in accordance to the measure of fire resistance required.

Capping and wall cavity installation Once above the area of wall where an accessory is to be sited, a cable may be dropped from the roof space in preparation for boxing out. This may be achieved in one of two ways: either by routeing the cables through the cavity or by running the installation on the fair face construction walls, protecting with plastic capping or oval conduit in readiness for

Figure 9.20. An emergency hole-maker for brickwork. The coupler acts as a renewable impact point.

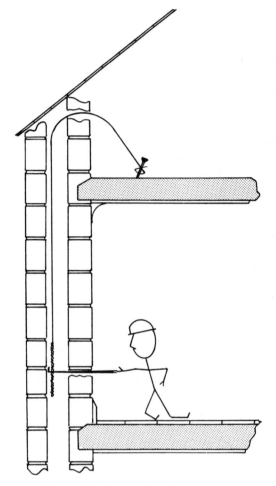

Figure 9.21. Cavity wall wiring techniques.

plastering. Should capping be decided, it is essential that the nails securing the material are galvanised. A good idea is to use 25 mm plaster-board nails as standard mild steel nails will often cause rust stains to appear under the finished plaster line. Plastic capping must always accommodate the full length of the cable drop from the ceiling to the top edge of the metal socket box. This will allow future rewires to be carried out without causing too much disturbance to the finished decor.

Capping may be trimmed to size quickly and neatly by using a sharp bolster and a medium sized hammer. This is ideally carried out on a concrete or metal base. Alternatively, a junior hack-saw will do the job.

Two lengths of twin or single width capping, secured together with PVC tape to form one large length, provides a simple means to span the underside of floor boards. Cables can be taped or tied on to the trailing end of the capping which may be pulled through with the minimum of trouble.

Should the cavity wall method appear more practical, drop a small plumb line from the loft to the room below. Mark both top and bottom with chalk and cut two 60 mm diameter circular holes on these impressions. This is obviously far easier when there is no ceiling. However, if there is a ceiling, a small locating hole can be made through

to the room below using a terminal screwdriver or a 2 mm drill. Next, tie approximately 100 mm of jack chain to a measured length of nylon cord and insert the jack chain into the top hole made in the roof space. The chain will fall through the cavity to the target area below and may be recovered by use of hooked fencing wire fished through the bottom hole. Once recovered, the nylon cord may be employed as a draw tape. This is usually a four-handed job, and Figure 9.21 will show how this may be carried out.

Thermal insulation Consideration *must* be given to cables grouped or installed in roof spaces and wall cavities which are served with thermal insulation.

Wherever possible, avoid cable contact with the insulation, but if direct contact is unavoidable a rating factor must be applied to the cable in order to reduce its current-carrying capacity. In practice this will mean using a larger sized cable if the circuit is designed to run to maximum capacity. Should the design current be low, no change is usually required in cable size. Details of rating factors may be found in the current edition of the *IEE Wiring Regulations* or *National Electrical Codes of Practice* and Appendix C of this book.

Boxing out Once all cables have been neatly laid and capping secured, the walls can be chopped out to accommodate the socket boxes. These should never be fixed back too far in the wall if using long machine screws is to be avoided. Ideally, allow 5 mm of box to protrude from the face of the unplastered wall so that the plaster line will be reasonably flush with the leading edge of the box as Figure 9.22 illustrates.

Always use two screws and expansion plugs in order to secure socket boxes to the wall. They are easily knocked out of square if nailed or fixed with one screw positioned in the centre of the box. A small, 80 mm water level is a useful addition to any tool kit and helps to maintain a good standard of box presentation when used to check socket box alignment.

When boxing out with foam insulated filled blocks, try cementing around the metal socket box in order to secure a fixing. Alternatively, use two

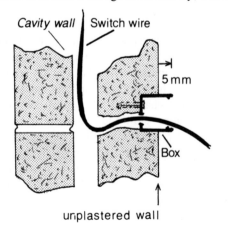

Figure 9.22. Allow 5 mm of steel box to protrude from the face of the unplastered wall.

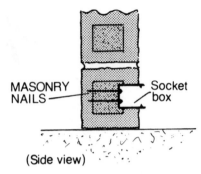

Figure 9.23. Long masonry nails often provide the way to secure pressed steel boxes into foam-filled thermal insulation blocks.

75 mm masonry nails to penetrate the foam insulation in order to pin to the unseen side of the block. Figure 9.23 helps to clarify this point.

Avoid installing boxes on damp walls in old houses. Dampness and general building sediment can often accumulate gradually over a period of time and can be the cause of nuisance tripping should a residual current device be fitted in circuit.

If the installation is designed so that boxes are fitted back to back, always fix one adjacent to the other as shown in Figure 9.24. Boxes placed in a back to back fashion may cause a weakness to develop within the wall resulting in the inability to fix the boxes in a conventional manner. If it is necessary to fit them in this fashion, render them in using a four to one mix of sand and cement.

Single socket boxes are usually manufactured with four tapped lugs; three fixed and one movable. Fit the box with the adjustable lug in the horizontal plane so a final alignment may be carried out when the accessory is wired and ready to fit.

Wiring from and through the ground floor ceiling space These days dwellings are often constructed with a solid ground floor. This provides no alternative other than to wire the complete ground floor installation from within the ceiling void serving the two floors. A modern three or four bedroomed house or flat could accommodate up to three final ring circuits in the design of the electrical installation. These would serve the needs of the first floor, ground floor and kitchen areas (Regulation 314 refers).

PLAN VIEW

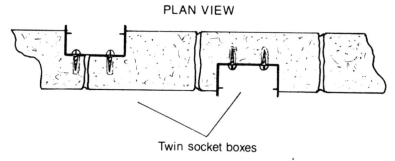

Twin socket boxes

Figure 9.24. Back to back socket outlets should be staggered.

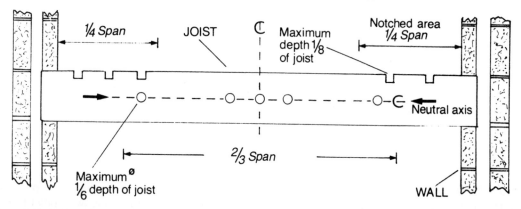

Figure 9.25. Holes and notches serving cable or conduit installations should only be cut in timber joists in specified regions.

To enable cables to be installed throughout the ceiling void, a series of holes must first be drilled. Unfortunately it still seems a common practice to drill holes and cut notches in timber joists wherever desired! Properly designed holes and notches will not seriously weaken the timber, but drilling and notching indiscriminately may cause structural problems. Figures 9.25 and 9.26 illustrate how holes and notches may be positioned in timber joists to accommodate cable and conduit installations.

Holes no more than one-sixth the depth of the joist may be drilled along the neutral axis. These must be made within the middle two-thirds span of the timber joist. Notches may be cut for conduit providing the joist is no less than 250 mm deep and is carrying a uniform load. These may be formed in the first quarter of the timber span measured from each supporting wall and must be no more than one-eighth of the depth of the joist.

A common mistake is often made by trying to force as many cables as possible into one hole.

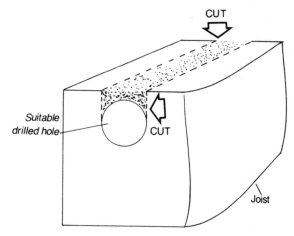

Figure 9.26. Section of a joist illustrating how a notch may be made in order to accommodate a conduit installation.

This practice can cause cable burn when additional cables are added and should be avoided. It is far better to drill a few more holes in order to maintain a good standard of workmanship.

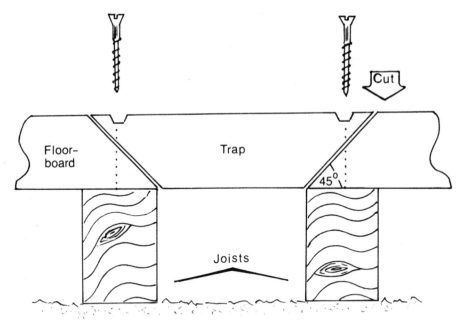

Figure 9.27. Cutting a small inspection trap in a conventional floorboard.

Floor traps Floor traps are often required for many different reasons in electrical installation work. Cutting a small trap in the middle of a long length of *in situ* floorboard may in itself be quite a daunting experience should you wish to cut the trap squarely to rest on the support timbers beneath. A quick and efficient way to overcome this problem is to cut, by means of a jigsaw, across the floorboards at an angle of 45 degrees, hugging the edge of the joist beneath. Repeat this procedure at the opposite end of the proposed trap and finally remove any interlocking tongues at the side of the trap by jigsawing between the boards. The removable trap may then be secured with four woodscrews driven in at 45 degrees in each corner as illustrated in Figure 9.27.

There are occasions, especially when rewiring occupied houses, when it may be difficult to lift floorboards in the conventional way. Take as examples the flooring serving a small cupboard where there are no butted flooring joins to work from or a passage way where the boards are laid at right angles to the run of the passage. Often it is difficult to select a suitable starting point in order to lift the boards required. If the width of the passage is wide enough it is sometimes possible to 'spring' the flooring upwards and to cut

directly over the joist. When this is not possible, another method is to run a jigsaw parallel to the joist, cutting the board twice across its width to form a trap. Watch for unexpected runs of cable or copper water pipes: these are extremely easy to damage when using a jigsaw.

As there will be no support to serve the newly formed trap, reinstating may be achieved by nailing a suitable size timber to the two joists beneath the level of the flooring as shown in Figure 9.28.

Kitchen installations

Potential current demand in a modern kitchen is usually sufficient to justify an independent final ring circuit. This may be installed in a similar manner to additional circuits serving other areas.

When wiring for socket outlets, cooker points and fused connection units, always use a water level to obtain a common height throughout the kitchen. Should tiles be placed around the walls, the fitted accessories will then appear completely uniform in relation to the tiles.

Kitchen installations are usually drawn separately, showing a detailed planned arrangement of the proposed kitchen layout. Always scale off

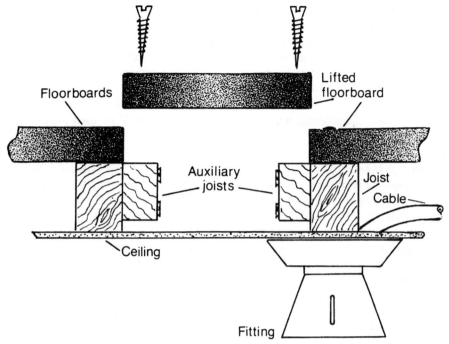

Figure 9.28. Auxiliary timbers secured to the side of the joist will provide support for the lifted floor trap.

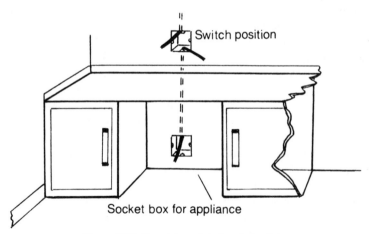

Figure 9.29. Remotely switched socket outlet.

the drawing to establish the exact area to be served by an electrical accessory. This way mistakes will not be made. Allow approximately 100 mm from the top of the worktop to the underside of the socket. This will allow adequate space for plugs and flexible cables.

Low level socket outlets serving refrigerators or waste disposal units generally need to be remotely switched from a position above the worktop. When this is required, wire the final ring circuit into a double pole 20 A switch. From this position drop a suitably sized cable to an unswitched socket outlet below the worktop as shown in Figure 9.29.

If a gas cooker is to be installed, remember to fit a fused connection unit at low level and adjacent to the appliance. This will provide a supply to serve any automatic ignition or electrical control circuit the gas cooker might have.

When a telephone point is required, use a standard 'first fix' 15 mm plaster depth box. Set the box into the wall a few millimetres to ensure the finished plaster line will be flush with the leading edge of the box.

Remember to carry out equipotential bonding in accordance with Regulations 413-02-01 to 03.

Nearing completion: first fix

Always label, by use of a permanent marker pen, cables needing to be identified so that others can take over in your absence. Trim back all cables to approximately 140 mm and tuck them securely into the socket boxes before plastering is carried out. This will help prevent the boxes being accidentally knocked out of square when the wall is plastered.

Check that all plastic capping or oval conduit is securely fixed and that each cable is completely covered. By doing this you will be helping not only the plasterers but also yourself.

Preparing for second fix

Once plastering has been completed and is reasonably dry, remove any contamination from the inside of each box and trim overhanging plaster to the outside edges. It is advisable at this stage to retap the threaded fixing lugs, as plaster and cement can cause difficulty, resulting in the brass machine screws being stripped of their threads.

Preparing PVC insulated and sheathed cables for connection may be made easier by the use of a sharp pair of side cutters and a cable stripper. Once the outer sheath has been removed, position the conductors and cut to suit the terminal positions within the accessory. This way, cable pinching and scuffing can be avoided. Too much cable accommodated in the box will cause problems when final fixing and commissioning is carried out. Strip off approximately 10 mm of insulation around the conductor. Try not to use pliers or side cutters as these can score the conductor, leading to the cable snapping off when pressure is applied by way of the terminal screw. Use a knife or cable strippers; it is far safer in the long run.

Lightly twist together all common conductors and screw them firmly home within the accessory. Remember to add a green and yellow striped oversleeve to protect and identify the bare current protective conductor (cpc), and loop a 'fly lead' from the back of the sunken box to the cpc terminal located on the rear of the socket outlet. Grommets should be used to protect the cable sheath.

If the fixing screws serving the accessory are too long, never cut to a required length using pliers or side cutters. Always use a threaded bolt shearer, which can often be found incorporated within a hand crimping tool. By using this method, the first few threads of the screw will be protected and fixing will be made less difficult.

It is easier to screw the accessory to the adjustable lug first. Final fixing may then be made without too much difficulty, especially if long screws are employed when boxes are placed further back into the wall than necessary.

Moulded accessories, once installed, give little trouble but there is a general need to protect brass electrical fittings from the effect of wall dampness once a room has been plastered. Dampness, reacting with the lacquered coating, causes the familiar blackening often found around the edges of brass accessories. This will happen if they are installed against new plaster, fair-faced brickwork or even freshly hung wallpaper. The effect is similar; only the length of time to achieve the unsightly discoloration is different. These days, manufacturers usually supply a moulded plastic gasket to place between the leading rear edge of the accessory and the wall. When the surface is completely dry the gasket may be removed without any further contamination appearing. To avoid unnecessary discoloration, always make sure that the gaskets are in place.

Brass and stainless steel accessories   Solid brass and stainless steel accessories, such as socket outlets, dimmer switches and fused connection units, can be ruined in seconds by a careless slip of the screwdriver during installation. To avoid needless frustration, stick four pieces of 20 or 25 mm wide PVC insulation tape around the accessory's fixing holes as shown in Figure 9.30

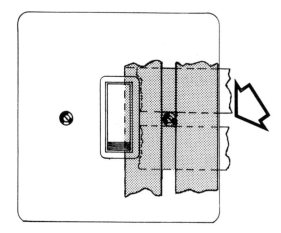

Figure 9.30. Brass or chrome accessories can be defaced at the slip of a screwdriver. Insulation tape will help safeguard against accidents.

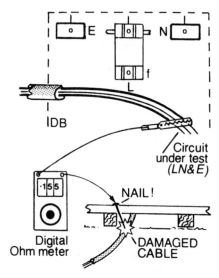

Figure 9.31. Nails driven through cables can be located by means of a digital ohmmeter.

and proceed with care. The tape may then be removed after the fitting has been safely installed. A small price to pay for peace of mind and a perfect finish.

Testing

Once the installation is complete all circuits must be tested in the manner prescribed within the current edition of the *IEE Wiring Regulations* or *National Electrical Codes of Practice*. In the UK this is carried out by measuring the total resistance, in ohms, between the phase/neutral conductor and current protective cable. The minimum value accepted is 500 000 Ω. If less, the circuit should be checked for pinched or scuffed cables or cables accidentally nailed creating a fault condition (see Chapter 14). Reference is made to Regulations 712 and 713.

An additional continuity test is then taken between all conductors as described in Chapter 6, under the heading 'Power circuit arrangements'.

Installation faults

Nails and fixing devices accidentally driven into cables during the period of installation can be very worrying for an electrician who has to find and then rectify the fault. A simple but effective way to locate a badly placed nail is to carry out tests using an insulation tester set to the middle range.

Replace one of the meter's service leads with a suitable length of insulated cable. Attach the free end of this supplementary lead to the phase, neutral and current protective conductor of the faulty cable. This is best carried out at the distribution board with the damaged circuit completely disconnected.

Having set the insulation tester to the 500 V scale, proceed to test all suspect nails using the original service lead. This can be seen more clearly by studying Figure 9.31. The nail responsible will clearly be identified when a 'zero' reading is displayed on the test meter. By switching to a lower voltage and to the ohms scale or by using a suitable continuity tester, it may be confirmed that the rogue nail has been located.

The cable may either be replaced or made good by means of a joint box in order to rectify the damaged circuit. After completion, retest to verify that the cable is in a satisfactory condition and reinstate in the distribution board.

Summary

1. Previous paragraphs have demonstrated the popularity of the final ring circuit together with several alternative lighting methods which are applied in electrical installation

work today. Although very different, all share similar practical installation techniques when PVC insulated and sheathed cable is used to form an installation.

2. Popular methods of wiring a lighting installation include:
   (a) joint box method;
   (b) loop in or three plate ceiling rose method;
   (c) central joint box system;
   (d) use of steel or plastic conduit.

3. All current protective cables are sleeved using green/yellow plastic oversleeving; Regulation 514-03-01.

4. A nylon or sprung steel draw tape can be used as an aid to draw cables through conduit with a greater efficiency.

5. Always install sufficient inspection boxes throughout the installation for ease of wiring or rewiring.

6. Conduit installations, although versatile, are very costly. Advantages include:
   (a) they are strong and sturdy;
   (b) they are easily rewired;
   (c) there is a choice of accessories to meet the requirements of the installation.

7. Confusion may be avoided by coding switch wires entering a multi-gang switch.

8. Volt drop must be considered arising from lengthy cable runs. If in doubt, use a larger sized cable or work out mathematically.

9. Start from the highest point of the project and work down.

10. Use galvanised nails when capping cables to avoid rust stains.

11. Apply a rating factor to cables which are in contact with thermal insulation or are grouped (see also Appendix C).

12. Metal 'knock-out' boxes serving switches and sockets should always be fixed using screws — *never nails*.

13. Timber joists should not be drilled or notched indiscriminately as this could cause structural weaknesses. If in doubt, consult an expert.

14. Add floor traps when using the joint box method of wiring.

15. A water level may be used to obtain a common height for electrical accessories when carrying out a kitchen installation.

16. A gasket placed between the rear leading edge of a brass accessory and a freshly plastered wall will prevent discoloration.

Handy hints

- Smear a little grease on bolts and screws serving outside or agricultural installations. This will help prevent corrosion.
- Never leave a hammer hooked on the upper rungs of a ladder. This is a dangerous practice which could lead to an accident.
- Always keep a first aid kit on site.

10 Steel and PVC conduit: installation techniques

In this chapter: Choice and size. Advantages and disadvantages. Pre-installation checks. Tools and equipment. Route planning and installation methods. Bending and setting solid drawn steel conduit. Gluing problems. Wiring complications. Types of cutting tools.

Early electrical installations were often carried out using split conduit to provide means for both mechanical and fault voltage protection. It was manufactured by rolling light gauge steel sheeting to form a tube; the sides of which were butted together throughout. Inspection boxes and bands were push-fitted over the conduit and made secure by means of small retaining screws. Practical difficulties experienced whilst forming split conduit to a required profile, forced operatives to rely heavily on available fittings. Although a reasonable aesthetic standard was accomplished, this method of wiring is never used now, but can often be found forming part of a redundant installation serving an old property.

Today we enjoy a wide choice of conduits from heavy gauge solid drawn to heavy gauge welded steel. Produced in a variety of dimensions, both categories are offered in four metric sizes (Regulation 521-04-01 i to iii): 16, 20, 25 and 32 mm (outside diameter). For specialised installations, three larger imperial sizes are available: $1\frac{1}{2}$, 2 and $2\frac{1}{2}$ inch diameter conduit.

Steel conduits are obtainable in a choice of two finishes:

1. Stove enamelled — for dry environmental use;
2. Hot dipped galvanised — for external use.

A wide selection of mild steel fittings are available in both stove enamel and galvanised finishes. Examples range from reducing sockets and couplings to locking rings and adaptors. Malleable cast iron, screwed circular inspection boxes are offered in a wide choice of design and include terminal, angle, through and three way. A selection of boxes most commonly used is illustrated in Figure 10.1.

Unplasticised polyvinyl chloride (PVCu) conduit was introduced to the industry after the Second World War. After a slow start it gained in popularity and is now widely accepted for both new and refurbishment work within the electrical installation industry. It is available in five metric sizes in black or white finishes: 16, 20, 25, 32 and 50 mm outside diameter.

A choice of light or heavy gauge high impact PVCu conduit is available to meet the demands of the installation. Light gauge is ideally suited for floor screeds or concealed work whereas heavy gauge is normally reserved for exposed situations.

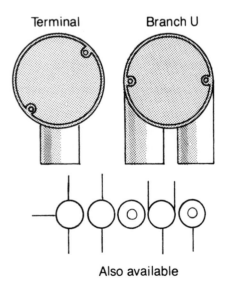

Figure 10.1. Standard conduit inspection boxes.

A variety of accessories and circular boxes are offered to meet the requirements of the task in hand. These may be used for either thickness of conduit. All fittings are secured by means of applying solvent to each surface to be joined. uPVC conduit installations will be debated more fully later in this chapter.

Oval conduit is seldom used to form an installation but is ideally suited as an inexpensive method of providing protection for cables buried in plastered walls. Usually supplied in 3 m lengths, oval tube is offered in a choice of five metric sizes: 13, 16, 20, 25 and 32 mm diameter.

Steel conduit

As with most electrical installations, there are both advantages and disadvantages to choosing steel conduit as a means of providing protection. A few of the more common aspects often referred to are listed; others will be debated in following paragraphs.

The advantages of steel conduit include the following:

1. May be rewired.
2. Can be added to.
3. When properly installed, steel conduit provides an excellent protective conductor.
4. Galvanised conduit is ideally suited for external use.
5. May be installed using mixed sizes of conduit.
6. Can be wired with a variety of cable sizes.

The disadvantages include the following:

1. High material and labour costs.
2. Additional man-hours required.
3. Careless or defective joins can cause large current drains should a fault condition occur.
4. Bushes and couplers can work loose during the lifetime of the installation. Maintenance is required.
5. If broken or damaged, steel conduit can be a less effective current protective conductor.
6. Stove enamelled conduit will corrode gradually if not properly maintained.
7. A high skill factor is required

Pre-installation check

Before any work is carried out it is advisable to calculate the number and size of cables required to be drawn through the conduit. Wiring Regulation 522-08-02 recommends that there should be adequate means of access for drawing cables both in and out. Remember to allow for any future additional wiring which could be undertaken after the installation has been completed. Consideration should also be given for both voltage drop and correction factors for cables grouped in conduit as advised in Regulation 525-01-01.

Tools and equipment

It is important to have a good basic selection of specialised tools. This way the task can be carried out with the maximum of efficiency and the minimum of difficulty.

Listed is a selection of tools and equipment required for this type of installation:

1. Stocks and dies of various sizes.
2. Conduit bender; complete with pipe vice.
3. Conduit formers for 20, 25, and 32 mm conduit.
4. Reamer (hand-held conical type).
5. Hack-saw (standard size).
6. Flat, medium cut file.
7. Retractable rule.
8. Soft leaded pencil.
9. Adjustable pipe grips.
10. Electric drill kit (110 V).
11. Draw-in tape and french chalk.
12. Paint brush and touch-up paint.

Try to keep tools and equipment in good working order. A pipe vice with blunt or worn jaws will allow the conduit to slip when thread cutting is carried out, consequently damaging the workpiece. Should this happen, remove the burrs with a flat file and paint the damaged area with a suitable protective paint to prevent corrosion setting in.

Screwdrivers with dulled or rounded blades can cause severe disfiguration to the heads of screws and can, in their own right, be potentially dangerous. Maintain all tools in good working order — be professional.

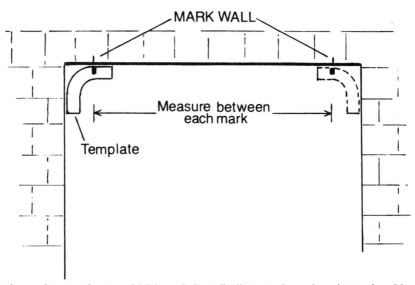

Figure 10.6. Offer the template to each corner. Lightly mark the wall adjacent to the mark on the template. Measure between each mark.

2. Using a soft pencil, indent the ceiling in both corners adjacent to the imprint on the template.
3. Measure between each ceiling mark.
4. Transfer the measurement gained to a suitable length of conduit and mark accordingly.
5. Align each mark to correspond to the rear of the former and use the machine to shape two perfect right angles.

Angles and sets may be formed in a similar fashion.

For each workpiece to match precisely with a purpose made template:

1. First bend the conduit to correspond with the angle of the template.
2. Remove the workpiece from the bending machine and position the template directly above it as though impersonating the finished set (Figure 10.7). This is best carried out by laying the unfinished work on a flat surface, such as a floor.
3. Use the template to achieve the required set and mark accordingly.
4. Transfer the prepared conduit to the rear of the bending former and shape the workpiece

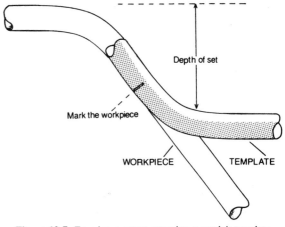

Figure 10.7. Forming a return set using a conduit template.

to bear comparison with the angle of the template.

With a little practice and determination, a good standard of workmanship can be achieved.

Galvanised conduit

Galvanised conduit is generally used for exterior installations but may also be laid in floor screeds and cast *in situ*. Whenever this installation practice is used, bends and sets should always be designed with generous radii for ease of wiring. Inspection

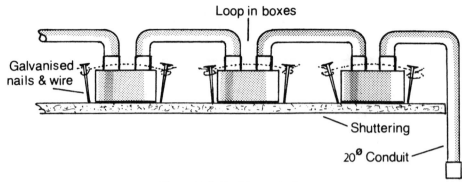

Figure 10.8. Conduit cast *in situ*.

and socket boxes should be packed with paper, with lids securely fixed to prevent concrete contamination filtering into the conduit. Complete installations cast *in situ* should be installed using galvanised loop in inspection boxes. Conduit, routed from box to box, is terminated in a vertical right angle and fixed to the construction shuttering timbers by means of a short length of light gauged fencing wire and galvanised nails. When the shuttering is removed the nails can be snapped off flush with the finished concrete slab and the paper withdrawn from the boxes. Figure 10.8 illustrates this principle.

It is a good idea to wear industrial gloves when handling conduit that has been stored outdoors in sub-zero temperatures. This will prevent skin damage to bare hands through sticking to the frozen metal. Galvanised conduit is ideally suited for external and environmentally damp installations but rubber gaskets should be used in conjunction with all inspection box covers to gain maximum protection from harsh weather conditions. Avoid the use of long machine screws for securing box covers. Should the cables be ranged in a manner which obstructs the course of the screw, biting or scuffing could occur, causing insulation damage leading to fault conditions. A smither of grease on the fixing screws will help prevent the steel lugs from corroding and allow the cover to be removed with ease.

Handy hints
- A wheel type pipe cutter, suitable for both PVCu and steel conduit, is an excellent alternative to a hack-saw.

- Use a heavy sprung steel or nylon draw-in tape to navigate through empty imperial sized conduits. Thinner counterparts often cause problems leading to unnecessary aggravation.
- An enclosed glass-fronted tungsten halogen floodlight is a good friend to have during the winter months but unprotected or open-fronted models should not be used in agricultural situations.

PVCu conduit

One of the main disadvantages of using steel conduit stems from the high cost of material requirements and the many man-hours required to complete an installation. Plastic, PVCu conduit, offers an alternative method reducing both material and labour costs.

There are many advantages, disadvantages and new skills to be gained from using this method of installation and it is this that will be debated within the next few pages.

PVCu conduit has many practical advantages over its steel counterpart. Less tools are required; it is easier and lighter to install and a junior hacksaw can often take the place of a larger one. Wheel-tape cutting tools can also be used to prepare the conduit if so desired.

Tools and equipment
As with steel conduit installation it is important to have the correct tools to maintain a good standard of workmanship. The basic tools required to work with PVCu conduit are as follows:

1. Junior hack-saw or wheel type cutter.
2. Hand-held reamer.
3. Retractable rule.
4. Electric drill set (110 V).
5. Assorted screwdrivers.
6. Draw tape.
7. Bending springs (20 and 25 mm diameter).

Conduit is formed by the introduction of a moulded steel spring into the workpiece and bending to a required shape. It is important to keep the spring in a good condition as damage can often cause the conduit to kink whilst being bent or set.

Select the size of conduit to be used and prepare the route in a similar manner described for steel installations. Because of its physical composition, PVCu conduit will soften and sag if exposed to heat. It is prudent to increase the frequency of fixings to approximately one every 900 mm should the installation be routed opposite skylights or through a naturally warm environment.

Heavy accessories should not be fitted to PVCu inspection boxes as the brass lugs have been known to pull away from the moulding when the ambient temperature is high. Warm weather can bring other problems, especially when long lengths of conduit are installed, for example, in warehouses or in agricultural situations. Small amounts of expansion throughout the length of the installation will compound, causing the conduit to bend and buckle. To reduce the effects of this phenomenon, expansion couplings (Figure 10.9) should be installed no less than 8 m apart. Be sure

to apply solvent only to the smallest section. The larger of the two sections is used solely as a guide so the expanded conduit can move freely within the coupling to avoid buckling. Reference can be made to Regulation 522-07-01.

Bends and sets: the effects of low temperature
As with solid drawn or welded steel conduit, do not form more than 180 degrees of bends or sets before adding an inspection box to the system. It is unwise to forge ahead without prior thought to potential practical wiring problems. Wiring is far easier when an installation has been carefully planned. Avoid making acute right angles in plastic conduit. It may look smart but it could cause difficulties when wiring.

Extremely low temperatures will cause PVCu conduit to snap when being formed. To overcome this problem, gently bring the temperature of the conduit to an acceptable level by use of an industrial hot air blower or similar source of heat.

Potential gluing problems Before solvent is applied to the workpiece, the area to be stuck should be thoroughly cleaned to avoid contamination. Dirty joints are often weak joints and can pull apart under stressful conditions such as when wiring is carried out or during the natural expansion and contraction cycle of the conduit. Apply the solvent to both surfaces to be bonded and twist the workpiece to spread the bonding agent evenly. Get into the habit of returning the lid to the solvent container after each use. If left off, sooner or later it will be knocked over, causing possible damage. This alone could prove to be a costly mistake.

Should you be without solvent, a temporary remedial joint may be made by wrapping one layer (one layer is usually all that is required) of PVC electrical tape around the conduit to be joined. Once pushed firmly home, the tape will provide a suitable remedial joint until it can be secured properly. Trim off any PVC tape which might be protruding.

Care should be taken using PVC solvent should plastic-lensed spectacles be worn. Accidental spillage or splashed solvent could ruin them in seconds. If in any doubt, always wear supplementary eye protection.

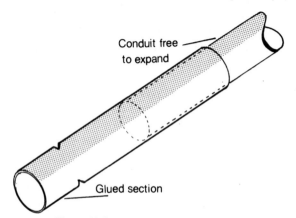

Conduit free to expand

Glued section

Figure 10.9. A PVCu expansion coupling.

Overcoming wiring difficulties

Providing the number of recommended bends and sets have not been exceeded in relation to the number of inspection boxes installed and the conduit is of sufficient size, then wiring will prove reasonably straightforward.

Once completed, make a final visual check to verify that the system is ready for wiring. One of the easiest of methods is to push the cables through the empty conduit. This will often be satisfactory, providing the installation is small and the amount of conductors to be dispensed is limited. By far the simplest method is to thread a nylon or steel draw-in tape through the conduit first, but even this can create local difficulties when negotiated throughout the system. If problems do arise, try feeding a hooked stranded conductor through the opposite end. When, by measurement, both draw-in tape and conductor are adjacent (Figure 10.10), twist the hooked conductor several times whilst gently removing it from the conduit. This is best achieved working in pairs; one pushing the draw-in tape, the other pulling the accompanying conductor. With a little patience, the draw-in tape can usually be removed successfully.

Cable preparation must be carried out thoroughly. Careless or clumsy methods of attaching conductors to the draw-in tape is often the recipe for disaster; especially where long runs are involved. An old and well-tried method is to

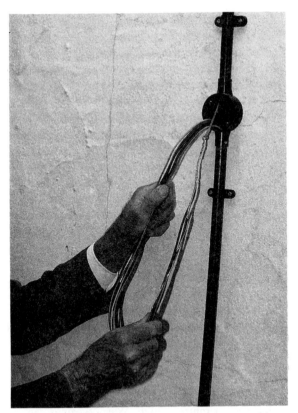

Figure 10.11. Preparing cables to be drawn into conduit.

bunch the cables together in a staggered configuration as illustrated in Figure 10.11. To prepare, remove approximately 80 mm of insulation from the end of each cable and twist the bared conductors around each other forming a spearhead of staggered cables. It is advisable to add a layer of PVC electrical tape around the spearhead. This will help prevent snagging occurring when the cables are drawn through the conduit. Once prepared, they can be firmly attached to the draw-in tape by way of the leading conductor.

Cable drums should be coupled to a suitable dispenser and wiring carried out by two people: one responsible for feeding the cables through the conduit whilst the other pulls. Cables should be served through the installation as straight as possible but should difficulties arise, sprinkle powered french chalk over the conductors. This remedial approach is noticeably beneficial and can be the solution to many a stubborn problem.

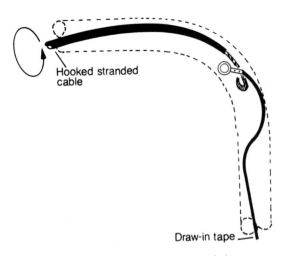

Hooked stranded cable

Draw-in tape

Figure 10.10. Draw-in tape recovery technique.

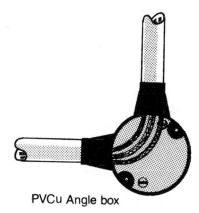

PVCu Angle box

Figure 10.12. Allow the cable's maximum radius (approximately 30 mm).

On long or complicated runs, wire one section of the installation at a time. Never attempt to negotiate the entire length of the system. Bends and sets will make it far too difficult. Cables routed through right angled and multiway inspection boxes should be allowed maximum radius as shown in Figure 10.12. Laying tight and hugging the inner periphery of an inspection box will add difficulties when additional wiring is carried out and could possibly obstruct the path of the box cover screw.

Socket outlets installed using steel conduit must be served with a supplementary current protective fly lead attached to the metal socket outlet box housing. This will provide protection should the socket be removed from the box. When an installation is installed using PVCu conduit and accessories, this practice is unnecessary.

Types of cutting tools

There are many different types of cutting tools to choose from; each having its own role to play in electrical installation engineering. Listed in random order are a selection of the more familiar tools used in our industry.

1. *Large hack-saw.* Ideal for cutting materials such as steel and plastic etc.
2. *Junior hack-saw.* Suitable for cutting materials where a large hack-saw would prove cumbersome.
3. *Jigsaw.* Used for cutting rectangular or oddly shaped holes in steel where conventional methods would prove difficult. A jigsaw may be used to cut timber but the blade must be changed to suit the texture of the material. Operatives who are unfamiliar with this type of tool should first be given a course of instruction before attempting to use the tool themselves.
4. *Hole saw.* Ideal tool for cutting holes in steel and plastic electrical inclosures. Clearance sizes: 16, 20, 25, 32, 38 and 50 mm diameter; generally made with 10 teeth per 25 mm. A 100 mm diameter masonry hole saw is also available for cutting walls and ceilings. Ideal for when extractor fans are to be installed.
5. *Files.* Used to provide a finish for steel and plastic work pieces such a trunking, conduit and tray work. Filing removes all burrs caused by cutting. Obtainable in flat, round, square, triangular and half round versions.
6. *Reaming bit.* An effective way to enlarge the bore of a hole. This tool can also be used for removing *burrs* from steel or plastic conduit caused through cutting. Some versions are equipped with a handle, others have to be fitted to a hand-operated drill.
7. *Stocks and dies.* These are used to cut and form threads on steel conduit. They are operated by hand in conjunction with a suitable cutting paste. Cutting sizes: 16, 20, 25 and 32 mm. Two imperial sizes are also available; $1\frac{1}{2}$ and 2 inches.
8. *Taps.* These are used to cut internal threads in holes slightly smaller in diameter than the diameter of the tap. Ideal for re-tapping threaded lugs serving pressed steel boxes or conduit fittings.
9. *Pad saw.* A disposal flexible blade firmly fixed into a permanent holder. This tool is ideal for cutting holes for recessed sockets and switches in dry-lining walls.
10. *Knife.* A tool for cutting or penetrating the insulation of cables. Ideally an electrician's knife should be of a type which has both a disposable and rectractable blade.

11. *Side cutters.* An ideal tool for cutting and indenting cables for future recognition when wire strippers are unavailable. Care should be taken when cutting small conductors as the copper waste has been known to fly and cause injury to an operative's eye.

12. *Wire cutters/compression tool.* Sometimes known as *service pliers*, this tool is designed for manual operation to remove the insulation and cut conductors up to 6 mm², shear small electrical screws or compress pre-insulated cable terminations. It is a versatile tool that should always be included in an electrician's kit, Figure 10.13.

13. *Rotary wood saw.* A hand-held machine, ideal for cutting *tongue and groove* floor boards. Can be mechanically set to cut wooden joists when preparing to lay conduit beneath the flooring. It is important that an operative is instructed how to use this machine before work is carried out.

14. *Drill bit.* This is the boring piece fitted to a drilling machine. High speed drill bits will serve most materials other than certain grades of stainless steel. Keep well sharpened at all times and rub over with light machine oil when storing. *Wood bits* and *augers* are used exclusively for timber and are used in conjunction with a carpenter's brace.

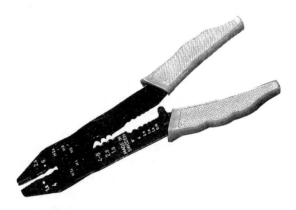

Figure 10.13. Service pliers. (Reproduced by kind permission of *Walsall Conduits Limited.*)

The importance of cutting accurately

It is important to cut accurately when undertaking any type of electrical installation engineering work. Holes wrongly placed create wasted time and additional work. Badly cut trunking can appear rather amateurish to the professional eye and will require remedial work before acceptance.

The way to gain perfection is by accurate measurement. Before cutting a hole, measure the position of the exact centre and draw a pencilled cross over the middle of the area to be worked. Should the pilot drill wander from the target it can be immediately corrected ensuring that the hole will be formed in the correct position.

Whenever trunking is cut it is important to remove waste material from the unwanted side of the workpiece. Accurate measurement and correct cutting techniques will always ensure good results. Never be tempted to guess!

Summary

Installation techniques overlap in many areas when PVCu or steel conduit systems are installed. The key points to remember when work is carried out using these methods of installation are listed.

1. Four sizes of PVCu conduit are available: 16, 20, 25 and 32 mm (outside diameter).
2. All sizes are obtainable in both light and heavy gauge PVCu conduit.
3. Four sizes of steel conduit are manufactured, together with three larger imperial sizes.
4. Determine both size and type of conduit needed to suit the requirements of the installation.
5. Keep a good selection of tools and plant to ensure that the task is carried out efficiently.
6. Plan both route and method of installation first.
7. Fix supporting accessories before undertaking any conduit work.
8. Cut holes and notches along the route wherever required.
9. Use a steel conduit template to assist with

the formation of difficult bends and sets.

10. Keep bending springs in a good working order. This will avoid damaging the conduit.

11. To avoid contamination, plastic conduit couplings should be cleaned before solvent is applied.

12. Always keep the lid securely on the tins of PVC adhesive when not in use.

13. Use a steel or nylon draw-in tape to assist wiring.

14. Attach cables to the draw-in tape in a staggered configuration for ease of wiring.

15. Wire in pairs, one operative feeding the cables into the conduit, the other pulling.

16. If difficulties arise or are expected, add a sprinkling of powdered french chalk over the cables.

17. Wire one section of the installation at a time.

18. Socket outlets and accessories installed within a steel conduit installation must be served with a flexible fly-lead attached from the accessory's earth terminal to an earthing pillar fixed to the rear of the metal box.

19. To prevent cable damage, avoid using long machine screws to secure conduit inspection box covers.

20. Rubber gaskets should always be used in conjuction with inspection box covers on all exterior installations.

Handy hints

- Condensation can be controlled by drilling a small drainage hole at the base of a low placed inspection box forming part of an outside installation. Regulation 522-03-02 refers.

- In the UK, conduit placed in a multi-service communal ducting should be identified with an orange band when required to be distinguished from other services. Regulation 514-02-01 and BS 1710: 1984 (1989).

- Warm to touch metal clad switches or socket outlets forming part of an outside installation could indicate that the accessory is contaminated with storm water. Once checked, drill a small hole at the base to allow future contamination to drain away freely. Reference is made to Regulation 522-03-02.

11 Mineral insulated and armoured PVC insulated cables

In this chapter: Cable, grades and sizes. Essential tools. Cable termination techniques. Installation methods. Dealing with older types of mineral insulated cable. Installation testing procedures.

Mineral insulated cables

Mineral insulated (MI) cables were first produced in 1936 and provide an ideal system of wiring for a wide variety of commercial and industrial installations. This highly reliable cable does not deteriorate with age, can be made moisture-resistant and is unaffected by sunlight. Its versatility may be shown by its ability to operate in temperatures of up to 523 Kelvin (approximately 250 °C) and will continue to function up to 1273 K (1000 °C) for shorter durations.

Conductors of solid drawn copper are inlaid into compressed powdered magnesium oxide insulation and encompassed within a soft copper sheath, as illustrated in Figure 11.1. Owing to its unique construction, mineral insulated cable has been known to be thoroughly flattened, yet still continue to function! The sheath provides both screening and mechanical protection and serves as a current protective conductor.

Grades and sizes

MI cable is available in a choice of three patterns, bare copper, PVC covered or LSF (low smoke and fumes) sheathed, of which several colours are obtainable to meet the demands of the installation. Plastic covered cable will withstand temperatures of up to 343 K, whereas LSF sheathing can contend with temperatures of up to 353 K (80 °C).

Manufactured in both light and heavy duty grades, the lighter grade cable is designed for voltages up to 500 V and may be selected in four sizes: 1.0, 1.5, 2.5 and 4 mm csa, in a choice of two, three, four or seven core cable. Cable of 4 mm size is usually only available with two current-carrying conductors.

Cables employed for heavy duty are designed to serve voltages up to 750 V and can be found in various sizes from 1.5 to 240 mm^2. A wide selection of cores is available but not all are seen to be represented in each of the 15 conductor sizes. There is, for example, just one size serving the 19 core cable, 1.5 mm^2 whereas a wider choice of seven sizes is available for the 3 core cable.

The unique construction and versatility of mineral insulated cable makes it ideally suited for wiring tasks demanding a high degree of reliability. It is often employed in fire alarm and emergency lighting circuits as well as boiler houses and environmentally hazardous installations. Reference is taken from Regulations 528-01-01 to 08.

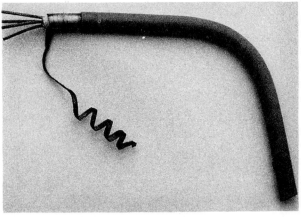

Figure 11.1. Mineral insulated cable.

Calculation of MI cable sizes The UK and all member European Community countries adopt a common criterion when standardising the size of all current-carrying conductors. This is determined by calculating the cross-sectional area of the conductor, not by use of its numerical standard wire gauge or diameter.

As an example, the diameter of a 1 mm^2 conductor has been calculated to be 1.128 152 1 mm and by applying the expression:

$$\text{cross-sectional area (csa)} = \pi r^2 \qquad [11.1]$$

where π is equal to 3.142 857 1 and
   r is the radius of the conductor in millimetres, the size of the conductor may be calculated. Substituting figures,

csa = 3.145 857 1 × 0.564 076 1^2
csa = 3.142 857 1 × 0.318 181 8
csa = 1 mm^2

Terminating tools

In order to terminate, a special brass gland and a screw-on pot is fitted to the cable (Figure 11.2). Once physically secured to the cable sheath, the pot is filled with a soft plastic sealing compound and mechanically sealed by means of a PVC cap.

As with most specialised cables, specific tools will be required for preparatory work and glanding to be carried out. A selection of the basic requirements follows:

1. Junior hack-saw.
2. Rotary cable stripper or side cutters.
3. Cable ringing tool — used in conjunction with side cutters to score an indent around the copper sheath to terminate stripping squarely. (This tool is not required when a rotary stripper is used.)
4. Pliers — used in conjunction with rotary stripper.
5. Pot wrench — used in conjunction with the gland to enable the pot to be screwed squarely on to the cable sheath (available in four sizes).
6. Hand crimper — used to crimp the plastic seal assembly to the pot (two sizes available for use with 20 and 25 mm pots).
7. Cable bender — to set or bend heavier grades of cable.
8. Roller straightener — used to iron out bent or crinkled cable.
9. Insulation tester.
10. MI cable gauge — for a quick check of cable size

Alternative tools

Should a rotary stripper be unavailable, side cutters or long nose pliers may be used to remove the required amount of copper sheath. As an alternative, an emergency tool can be constructed from a 6–8 mm diameter circular steel rod. An old poker with the point removed would make an excellent remedial tool. To make and commission:

1. First cut a 20 mm slot across the face end of the rod (Figure 11.3(a)). This is best carried out using two hack-saw blades in tandem.
2. Prepare approximately 20 mm of torn copper sheathing by flattening and placing into the slotted rod.
3. The copper insulation may then be removed as though opening a tin of sardines by rotating the tool and circumnavigating around

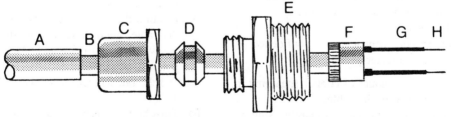

Figure 11.2. Mineral insulated cable termination: A, PVC outer cover; B, copper cable sheath; C, compression gland; D, compression olive; E, gland housing; F, brass screw-on pot; G, conductor insulating sleeving; H, conductor. (The brass screw-on pot houses both sealing compound and crimping cap.)

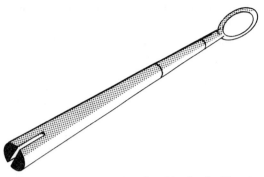

Figure 11.3(a). An emergency mineral insulated cable stripper.

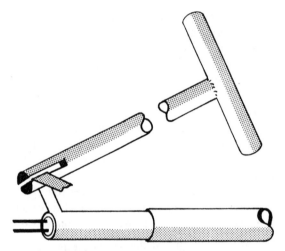

Figure 11.3(b). Place the prepared flattened copper sheathing into the slotted section of the rod.

the periphery of the cable as shown in Figure 11.3(b).

Cable termination techniques

Terminating mineral insulated cable can be quite a daunting task for an operative who is not familiar with the procedure. A step by step approach is given which, when taken in numerical order, will help the reader to build to a reasonable professional standard.

Remember: practice contributes to confidence — confidence leads to greater proficiency!

1. Calculate the amount of copper sheathing to be removed from the cable.
2. Trim the end of the cable squarely using a junior hack-saw.
3. Offer a rotary stripper to the end of the

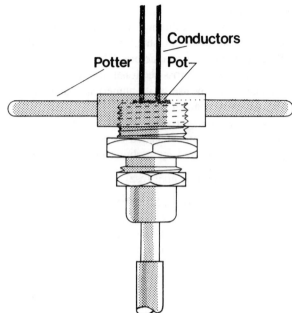

Figure 11.4. Pot ready to be screwed on to the cable.

cable and make any necessary adjustments to accommodate the outside diameter of the cable sheath.

4. Rotate the tool in a clockwise direction, slightly pushing forward to remove the copper sheath.
5. When the required amount of sheathing has been removed, clamp a pair of pliers to the cable sheath to act as a stopping mechanism. This will also provide a square leading edge between the sheath and the exposed copper conductors.
6. Carefully remove the tool from the cable.
7. Offer the plastic shroud and brass gland to the cable.
8. Place the pot squarely on the end of the prepared cable.
9. Position the gland, now free to move along the cable, gently into the pot and screw the pot-wrench onto the gland. At this stage the pot will be sandwiched between the gland and the pot-wrench as shown in Figure 11.4.
10. Turn both wrench, pot and gland with sufficient pressure to attach the pot securely to the cable.

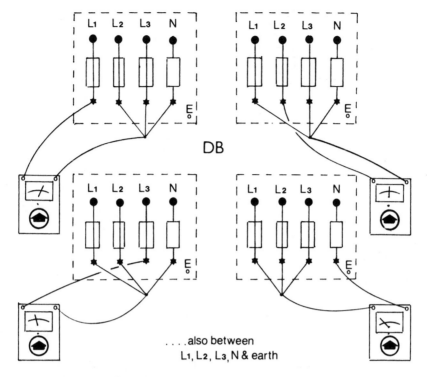

Figure 11.10. Insulation testing a three phase and neutral installation.

Larger installations supporting, for example, 200 outlets, each with an average installation resistance value of 50 MΩ, would show an unsatisfactory value when tested at the distribution centre. This is because each outlet is wired in parallel formation with the principal supply conductors and therefore reduces the collective value of the installation (see Expression [2.18]). On larger systems, subdivide the installation into sections of no less than 50 points for the purpose of testing. This will help to increase the underlying value of the system and avoid unacceptably low insulation test results.

Figure 11.11 defines the term 'points' in graphical form. Each switch, socket outlet and luminaire installed as part of an installation is regarded as a point. Accessories which are integrally switched can be considered as a single point.

Mineral insulated cable installations can often prove difficult for an inexperienced operative. Practical experience should be sought once the theoretical concepts have been mastered.

Handy hints

- Nails should be removed from floorboards when they are lifted in order to minimise the risk of accidents.
- Accurately positioned holes can be drilled by first placing a cross over the position to be worked. Errors can then be immediately corrected.
- Bonding clamps should be fitted only around copper pipes which are filled with water. When empty, they can easily be crushed or distorted.

Armoured PVC insulated cables

Armoured PVC insulated cable is of robust construction and ideally suited for power transmission and control circuits. Known also as stranded wire armoured cable (SWA cable), it can be made secure by means of purpose formed cleats or, when grouped, by use of metal strapping. Alternatively, it may be laid directly in the ground

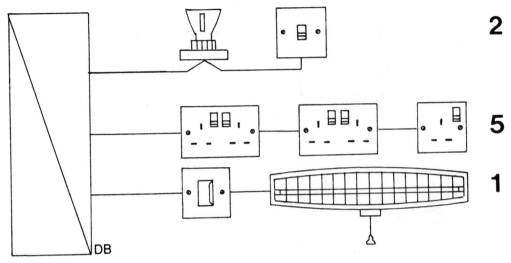

Figure 11.11. A twin and single socket outlet would be regarded as three outlets whereas a heater served by a fused connection unit, just one.

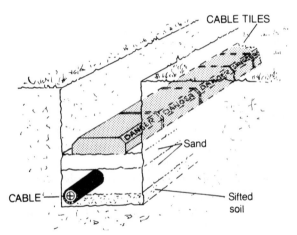

Figure 11.12. Armoured PVC insulated cables may be laid directly in the ground.

as illustrated in Figure 11.12 or placed within a subterranean ducting. Regulation 522-06-03 confirms.

Armoured PVC insulated cable is terminated by use of a purpose made gland acting as an anchorage point for the steel wire armouring sandwiched between its housing. This will be examined in detail later.

Smaller cables are manufactured in a number of sizes from 1.5 to 16 mm in cross-sectional area. A selection of between 2 and 40 cores is available, but not all cores are represented in each of the

conductor sizes. Larger cables are made in a choice of 2, 3 and 4 cores with cross-sectional areas from 25 to 400 mm.

The cable is constructed using solid or stranded copper conductors protected with coloured or plain numbered PVC insulation. A thin PVC bedding sheath wrapped around all conductors provides support for a single layer of galvanised steel wire armouring. This may act as a current protective conductor, if so required. Alternatively, a spare core may be colour-coded and used for the same purpose. Finally the armouring is protected with a thick moulding of PVC acting as an outer sheath as shown in Figure 11.13.

Installation techniques

It is most important to have the correct tools and equipment whenever an armoured cable installation is being carried out. Without these, the task can become difficult and possibly hazardous.

Cable can be dispensed from the master drum in a safe and efficient manner by use of a pair of cable jacks and a solid circular steel bar. The drum is elevated about 90 mm above ground level as shown in Figure 11.14 and the cable drawn from the top. It is important that it is never left unattended and free to rotate. By manhandling the drum, whilst others are pulling in unison, greater control may be gained dispensing in this fashion. It

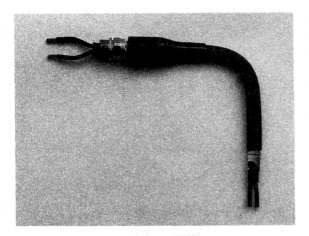

Figure 11.13. An armoured PVC insulated cable.

often helps to smear a little grease on the axle to reduce the resistance generated by the weight of the drum. This will make it far easier to control.

To avoid discharging vast quantities of cable in one operation the cable can be 'snaked' as illustrated in Figure 11.15. This way only a small amount of the total length is pulled at a time which consequently reduces the work load. Another way is to position the drum at a mid-point position and undertake one half at a time. The cable remaining can then be removed from the drum and laid in a snaked configuration as shown in Figures 11.15 and 11.16. To do this successfully, plenty of room will be required. Thus this technique would be unsuitable for certain types of installation.

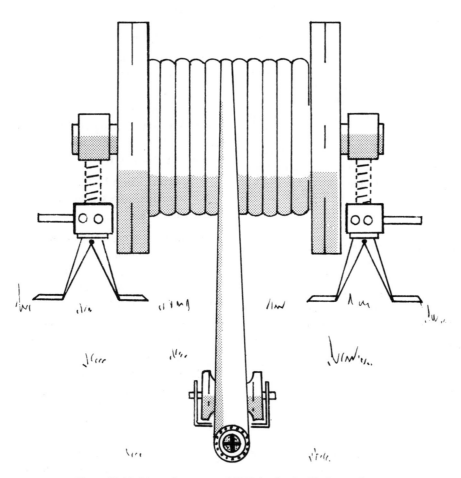

Figure 11.14. Dispensing armoured PVC insulated cable from a drum.

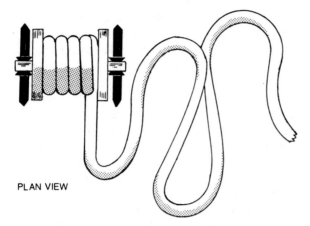

PLAN VIEW

Figure 11.15. Try 'snaking' the cable in order to make lighter work of the task in hand.

Through manhandling it can be easy to acquire kinks and bends laying opposite to the natural twist of the armouring, making the cable difficult to lay and conform with expected standards. Always monitor and check the dispensed grounded run to avoid unnecessary work at a later stage. As an additional aid, cable rollers may be used but this method is usually only practical when long straight runs are available.

Heavy vehicles and plant should be discouraged from driving over armoured cable especially when laid over rough or unfinished surfaces. This will reduce the risk of deformation and flint stone penetration.

Catenary suspension (reference is made to Regulation 521-01-01) Armoured cable may be safely suspended by use of an overhead catenary wire providing the size and type used will adequately accommodate the total weight distributed. It may be attached by means of heavy duty plastic or nylon cable ties placed at intervals of approximately 250 mm as illustrated in Figure 11.17. This is best carried out when the cable and straining wire are lying on the ground. Lifting could prove to be a heavy undertaking and mechanical assistance may be required in order to carry out the task safely.

When completed, both ends of the catenary wire should be securely anchored to a suitable ceramic insulator.

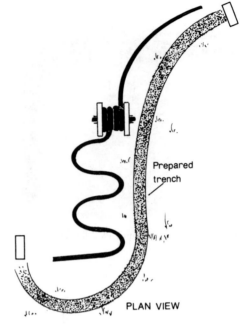

Prepared trench

PLAN VIEW

Figure 11.16. Another way to reduce the weight of cable pulled is to position the full cable drum at a mid-point position.

Tools and terminations

Unlike MI cable, armoured PVC insulated cable requires the minimum of tools to prepare a satisfactory cable termination:

1. A strong sharp knife.
2. Two adjustable spanners.
3. A hack-saw.

Terminations are usually straightforward to assemble and seldom present practical difficulties — unless the wrong sized gland is used! The following lists 12 easy to follow guidelines to consider whenever glanding is carried out:

1. Calculate, by measurement, the amount of armouring to be stripped, and mark accordingly.
2. Saw around the periphery of the cable at the marked position cutting the outer PVC insulation whilst scoring approximately 50 per cent of the galvanised armouring beneath it.
3. Remove the outer plastic sheath using a sharp strong knife from the point of termination.

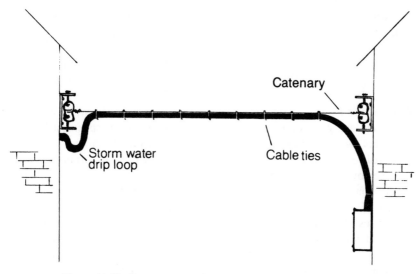

Figure 11.17. Catenary suspension: armoured PVC insulated cable.

4. Displace the galvanised armouring by untwisting three or four strands at a time. Providing the scored periphery is sufficiently deep the armouring will snap off both neatly and squarely.
5. Slide the plastic shroud onto the cable beyond the point of termination.
6. Remove sufficient outer sheath to enable the armouring to be accommodated within the conical section of the gland (Figure 11.18).
7. Slide the gland nut section onto the cable, placing it below the plastic shroud.
8. Splay open, in conical fashion, the short length of prepared armouring as described in Item 6 and Figure 11.19.
9. Offer the conical section of the gland to the splayed armouring.
10. Reunite the two sections of the gland and tighten as illustrated in Figure 11.19. At this stage the armouring will be sandwiched between the conical section and top locking component of the gland as described in item 7.
11. Using a sharp knife, remove the inner bedding sheath to a point approximately 10 mm from the face of the gland. Care must be taken not to damage the insulation covering the conductors by cutting too deeply.

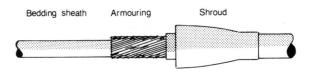

Figure 11.18. Exposing the wire armouring.

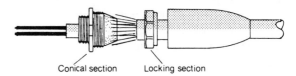

Figure 11.19. Tighten the locking section after the splayed armouring has been offered to the conical section of the gland.

12. Place the plastic shroud over the gland and test using an insulation tester set at twice the nominal voltage of the installation. Discharge any capacitance within the cable after completing the tests.

It might seem difficult at first sight, but once one or two compression glands have been successfully mastered, terminating armoured PVC insulated cable will seem less of a formidable task.

Summary

1. Mineral insulated cable does not deteriorate with age. It is ideally suited for

discriminatory and critical circuitry or where high ambient temperatures are present.

2. The cable is obtainable in two grades: light duty for voltages up to 500 V, heavy duty for voltages up to 750 V.

3. Cable terminations are made by removing the outer sheath and applying a purpose made gland and pot which is then permanently sealed.

4. A synthetic rubber oversleeving is then placed over both conductors and wedged into the sealing cap.

5. Terminations are tested by use of an insulation tester set for twice the nominal working voltage of the installation.

6. Think ahead when planning cable runs.

7. Three choices of high temperature seals are available to suit the conditions of the installation.

8. Test thoroughly when completed. Conductors can be identified by means of a continuity test meter or by using two telephone hand sets wired in series formation as shown in Figure 11.7.

9. On larger systems, subdivide the installation into sections of no less than 50 points for the purpose of testing. This will avoid unacceptably low insulation resistance values.

10. Connect both phase and neutral supply conductors together and test between them and the principal current protective conductor. A further test should be taken between each current-carrying conductor. Test results obtained should not be less than 0.5 MΩ.

11. Armoured PVC is an ideal cable suited for power transmission and control circuits. Available in a choice of sizes and various cores.

12. Avoid kinks when laying out.

13. Install a little at a time using cable rollers (if required). Use the correct plant.

14. Terminations may be prepared with the minimum of specialised tools.

15. Measure the amount of cable to be removed to enable the termination to be accurately offered to the accessory served.

16. Observe the guidelines offered when cable termination is carried out.

17. Once complete, test at twice the working voltage of the installation.

18. After testing, short circuit each core to each other and each core to the armoured sheath to eliminate stored capacitance.

Handy hints

- Removing the outer insulation from armoured PVC insulated cable during severe winter conditions can be difficult and often hazardous. By use of a hand-held electric hot air blower the plastic cable sheath can be gently warmed to allow safer cutting.

- A brass earthing tag sandwiched between an armoured cable gland and the equipment served will allow the attachment of an external supplementary current protective conductor.

- It is wise to lap a few turns of coloured electrical tape around your tools. This will not prevent them being stolen but it will provide a means of recognition should they be left unattended or inadvertently placed in someone else's tool box!

12 Domestic immersion heater and central heating control circuits

In this chapter: Immersion heater circuitry, fault-finding and element changing. Overtight elements. Bottom entry storage vessels and limescale contamination. Rewiring. Off-peak installations. Central heating control circuit wiring techniques. Cable terminations and general fault-finding.

Modern domestic immersion heater cylinders are constructed offering a wide diversity of hot water storage capacities ranging from 115 to 210 litres. Moulded foam thermal insulation can be added to the outer surface during the manufacturing process. Immersion elements, designed for domestic use are generally rated at 3000 W single phase. Industrial and agricultural models are produced in a wider range of power ratings and may be selected for use on either single or triple phase circuits. Chapter 12 has been written in accordance with the requirements of Regulations 554-05-01 to 03.

Designing the circuit

Figure 12.1 illustrates a typical immersion heater circuit terminated by means of a double pole (DP) switch with a pilot lamp. It is tempting to originate the circuit from a fused connection unit forming part of a final ring circuit when time is at a premium — but this method should be considered as bad practice. Current flowing through the heater element would greatly reduce the potential current that could be drawn from the final ring circuit and this would obviously lead to overcurrent problems.

Wiring from the control switch to the immersion element should always be installed with heat-resistant flexible cable. Standard PVC or rubber insulated cables will deteriorate and become hard and brittle when exposed to heat generated by the element and hot water system.

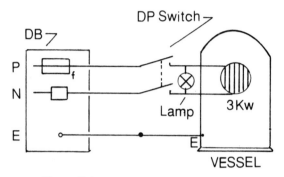

Figure 12.1. Typical immersion heater circuit.

Fault diagnostics

In hard water areas limescale depositis will attack the sides of the immersion heater element and gradually, over a period of time, cause microscopic pin holes to appear. Should the installation be served with a residual current device, tripping will occur. To remedy this, the element must be removed from the vessel and a new one fitted.

A faulty immersion heater forming part of an installation served by undersized mains can cause flickering or undulating patterns to occur in the supply voltage. This is more apparent when the lights are on and the mains are served by a TN-CS or TN-S supply system (see Chapter 8).

An element can often become faulty through age. Testing should be carried out using a standard multimeter switched to the ohms scale. By applying Expression [2.2] ($W = V^2/R$), the resistance of the element may be quickly calculated before any instrumentation test is implemented. With the circuit completely isolated a test should be made between the phase and neutral terminals serving the element and the value in ohms noted. If the element is sound, the value in ohms obtained

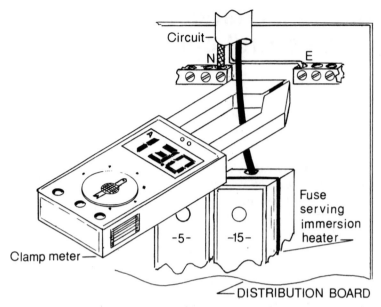

Figure 12.2. Clamp meter method of testing an immersion heater element.

should compare closely to the figure calculated before the test.

Check the insulation value of the element by measuring the total resistance in ohms between both current-carrying terminals and earth. This is best achieved using a standard insulation test meter where a fault condition will appear as a value approaching zero ohms.

As an alternative method, a clamp meter may be used to confirm a suspected internal open circuit. Testing is ideally carried out at the distribution centre where conditions are less cramped by adopting the following procedure. First check that the immersion heater thermostat is working and is calling for heat. Next, carefully remove the cover of the distribution centre and verify that the respective overcurrent device is functional and properly rated. Switch the clamp meter to a suitable scale (say 15 amp full scale deflection) and open the jaws to encompass the phase conductor serving the immersion element, Figure 12.2. If the element is sound and working as designed, a value of approximately 13.0 amps will be recorded by the instrument. Conversely, an extremely low current value or no reading would indicate an integral open circuit or fault condition within the heating element.

Intermittent faults Earth leakage problems may be adequately dealt with using an insulation test meter ranged to 500 V. Sometimes an element will develop an intermittent fault to earth, causing temporary power loss when served by a residual current device. Often the only way to identify this problem is to use a much higher test voltage, but when doing so, to isolate the element physically from the installation. This will prevent damage to domestic or commercial electronic accessories.

Changing an immersion heater element

Hot water storage vessels are generally installed in airing cupboards. Before work is carried out, the contents should be removed and the floor covered with old newspaper to soak up any water spilt accidentally. Once the water supply to the cylinder has been turned off, isolate electrically.

Careless removal of the element may result in permanent damage to the cylinder. A specially designed box or ring spanner should always be used, placing equal pressure on either side of the tommy-bar whilst turning. Care should be taken to keep the box spanner squarely aligned to the element flange since slipping could severely impair

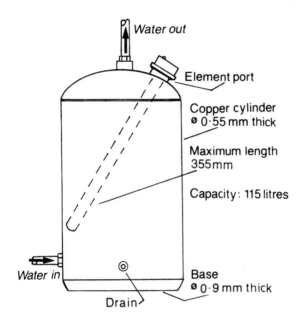

Figure 12.3. Hot water storage vessel; design features.

the vessel by buckling or rupturing the thin copper wall.

Figure 12.3 illustrates the basic design features incorporated into a modern hot water storage vessel.

Once the element has been removed, thoroughly displace any limescale deposits around and on the face of the threaded flange. This will help to maintain a watertight joint when the new element and gasket are fitted. Never leave the old fibre gasket attached to the neck of the vessel. If overlooked or left intentionally it could lead to a leakage of water at the neck of the element port. It is helpful to deposit a smear of jointing paste over the threads and gasket of the new element. This will help to secure a watertight joint. Should the cylinder still leak, a small quantity of hemp wound round the threads of the element will generally solve the problem.

Overtight elements
Overtight elements can often cause difficulties when failing to respond to conventional methods of removal. A gentle flame played around the neck of the element will often be sufficient to weaken the hold before withdrawing in the customary manner.

This method is not practical if the copper storage vessel is factory clad in thermal insulation.

Problem stopcocks
If, after withdrawing the element, it is found that the stopcock is not holding and water is still entering the vessel, siphon off sufficient water to enable the replacement to be fitted without risk of spillage. The reduced water level will allow work to be carried out efficiently and will provide access to the inner threads so that descaling or retapping may be carried out. Scale contamination accumulating at the base of the element port can also be attended to at this stage.

Bottom entry storage vessels

Generally there is considerably more work involved when a replacement element is required for this type of storage vessel:

1. First electrically isolate and disconnect the flexible service lead to the element.
2. With the vessel still filled with water, release the element by turning anti-clockwise for approximately half a turn.
3. Attach a convenient length of hosepipe securely to the storage vessel drain cock outlet.
4. Release the drain cock valve.

This method will reduce the risk of damage to an empty vessel should the element prove stubborn or difficult to handle.

Whenever heavy limescale deposits prevent emptying by means of the drain cock, controlled measures of water may be dispensed through the element port into a suitable receptacle. This is a painfully drawn out method of emptying a bottom entry hot water storage vessel but never try to increase the flow of water as an unscrewed element will cause flood conditions in a matter of seconds. Patience and a little assistance make a good recipe for success when draining a storage tank in this fashion.

Limescale contamination
One of the advantages of choosing a bottom entry hot water vessel is that accumulated limescale may

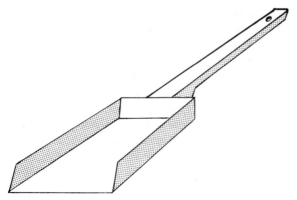

Figure 12.4. A simple scoop, suitable for removing limescale from a bottom entry water storage vessel, may be easily made in the workshop.

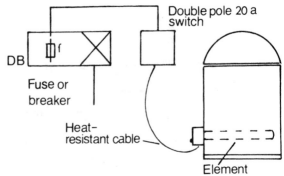

Figure 12.5. Immersion element: basic wiring arrangements.

be successfully removed with the aid of a prefabricated or home-made scoop. The scoop (Figure 12.4) may be employed whenever an element requires changing. This will help to maintain the interior in a reasonable condition. It can be tempting to leave the deposited limescale, but to do the job professionally, all contamination should be removed. A top entry storage vessel can compound the problem, making the task virtually impossible to carry out; with a little patience, most of the accumulated limescale may be successfully extracted from a bottom entry vessel.

A severely contaminated tank can cause difficulties when a replacement element has to be fitted. Accumulative limescale can completely engulf a heating element, rendering it impossible to remove. Under these conditions it would be prudent to seek advice from a plumber.

Rewiring

Replacement flexible cable should be of the heat-resistant type and appropriately sized to suit the load of the appliance. Figure 12.5 illustrates the relationship between the heating element, thermostat and service cable.

Over a period of time, copper heating element terminations will deteriorate in strength and soften, so care should be taken to support the terminal posts when rewiring. This will avoid accidentally twisting them off.

Off-peak installations and wiring techniques

Figure 12.6 depicts a typical dual-element hot water storage vessel. The day tariff arrangement occupies the upper section, the lower portion being exclusively reserved for the off-peak installation. This is wired in similar fashion to the day tariff arrangements except that the supply is taken from a local off-peak distribution centre. Wiring from a local isolating switch is carried out in the usual manner using a suitably sized three core flexible heat-resistant cable.

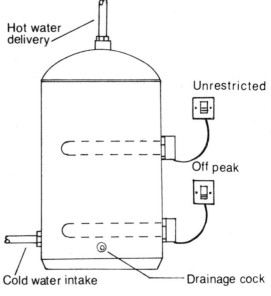

Figure 12.6. A hot water storage vessel incorporating both unrestricted and off-peak tariff heating elements.

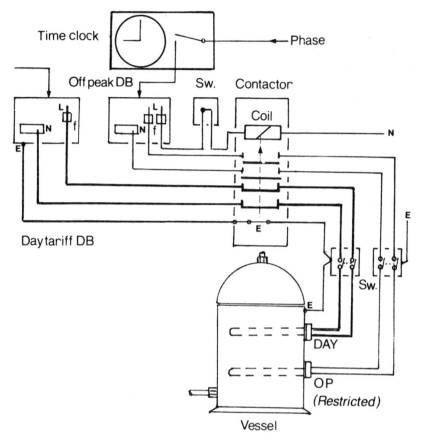

Figure 12.7. A contactor may be employed to serve both unrestricted and off-peak tariffs.

Alternatively, both day and off-peak (OP) circuits can be controlled automatically by means of a suitable changeover contactor. This is shown in schematic form in Figure 12.7. The de-energised contactor provides an uninterrupted supply to the day tariff immersion element. When the timed off-peak installation is activated, the contactor is instantly energised, automatically providing switching facilities from day to off-peak in one operation.

Bonding
Whenever rewiring is carried out, bear in mind that Regulation 554-05-02 calls for an independent current protective conductor to be connected from the main earthing terminal to the water pipe through which water to the vessel is provided.

Handy hints
- A smear of petroleum jelly or grease over the threads of a PVC solvent tin will prevent the lid from sticking and provide easy access.
- To avoid accidental spillage and possible contamination, PVC solvent should be made secure when not in use.

Domestic central heating: wiring and controls

There seems to be an underlying but understandable reluctance among less experienced operatives to undertake domestic central heating control wiring. Faults can appear complex and compounded should the system be installed incorrectly, but generally the task may be carried out with the minimum of trouble in a relaxed routine manner provided work is implemented strictly to the manufacturer's wiring instructions and all cabling is clearly identified for final connection.

Control circuitry serving domestic central heating systems may be installed using the following methods:

1. Direct wiring method to central programmer.
2. Central joint box method.

The direct wiring method

The majority of basic programmers are designed taking the form of an electronic timing device controlling two or more intrinsic relays. These are internally wired to serve, automatically or manually, heating controls and current-consuming equipment such as diverter valves and circulation pumps, the cylinder thermostat and boiler; all of which are supplied from the central programmer.

Interconnections are made within the programmer's terminal housing in strict accordance with the manufacturer's wiring instructions.

Figure 12.8 illustrates a basic central heating gravity system. Boiler, circulating pump and room thermostat are controlled by a central programmer.

Programmers are usually equipped with a battery back up facility enabling the electronic timer to function up to 24 hours after power failure. This, of course, will automatically recharge once power has been restored but it is advisable to replace the batteries once every four or five years to ensure reliability.

Wiring techniques Wiring should be carried out using a lighting grade PVC insulated and sheathed cable. Larger sizes might, collectively, be too bulky to be physically accommodated within the programmer.

The supply may be taken either from a fused

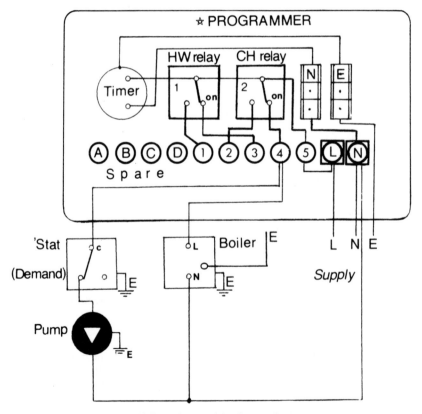

Figure 12.8. Basic central heating gravity system.

connection unit forming part of a final ring circuit or be provided independently from a distribution centre. In order to ensure local isolation, the circuit should be terminated into a 20 A double pole switch. Whichever way is chosen, the circuit protective fuse or breaker should not normally exceed 6 A.

Choose carefully where the programmer is to be positioned. Many surface models are designed with top fixing screws, so enough headroom should be left to allow for final assembly after wiring and connecting has been completed.

Room thermostats are preferably sited on a north-facing wall approximately 1.5 m from the finished floor level. To work efficiently the thermostat must be kept free from direct sunlight and uninfluenced by external heat sources. In-line central heating and hot water controls are usually supplied and fitted by a plumber.

More elaborate wiring arrangements serving cylinder thermostats and mid-position diverter valves need be no more demanding to install than a basic gravity system, other than additional cables to consider and extra conductors to terminate at the programmer. Figure 12.9 shows a typical arrangement controlling a fully pumped system. Remember to bond the central heating system in

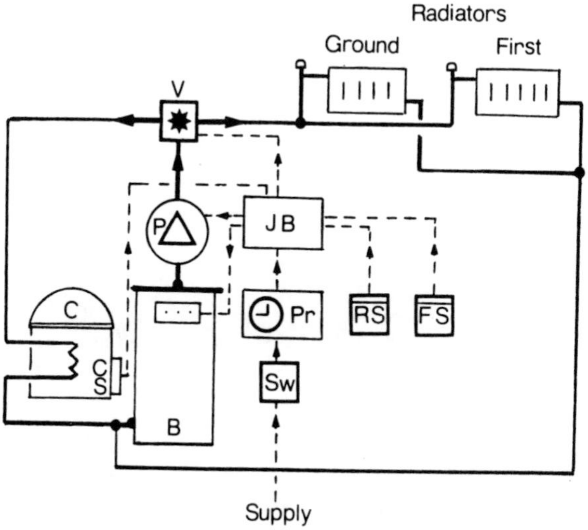

accordance with Regulation 413-02-02 (iv) and Chapter 8 of this book.

Cable installation and termination guidelines Adopt a method which will avoid mistakes from being made and consider the following listed guidelines:

1. Identify the roll of each cable at either end using a permanent marker pen.
2. Connect with care. One wrongly terminated conductor could produce a multitude of fault conditions.
3. Check the wiring as the installation progresses. It is easy to make mistakes with so many cables playing different roles.
4. Strip one cable at a time and screw the prepared conductors firmly home. Loose terminals will lead to arcing within the terminal housing causing damage to the programmer.
5. Add cable identification markers to each conductor to correspond with terminals occupied. Regulation 712-01-03 (ii) confirms.

Many programmers have several spare terminals incorporated within the main terminal housing. These may be used for interconnection purposes and are often identified with letters to differentiate them from functional terminals.

Fault-finding When fault conditions arise as a direct result from wiring errors, the system should be tested using a good quality multimeter. The use of a neon test screwdriver or similar electronic voltage sensitive device should be avoided. Although extremely helpful and a practical aid, they can respond to electromagnetic induction induced into a recognised 'dead' conductor. In dry environmental conditions it is possible to generate an illuminated display by electricity propagated from nylon carpets!

Electronic circuitry now plays an ever-increasing role in today's technology and severe damage can result from testing with a high voltage insulation tester. Disconnect all electronic equipment before checking the installation and make a written note where the conductors are terminated within the appliance. A wrongly placed wire can easily lead to prolonged fault conditions. Unfamiliarity with

Figure 12.9. Typical central heating control system: B, boiler; C, cylinder; CS, cylinder 'stat; FS, frost 'stat; JB, central joint box; P, pump; Pr, programmer; RS, room 'stat; SW, double pole switch; V, valve.

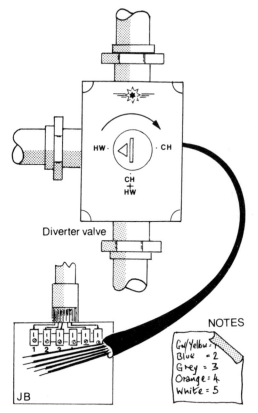

Figure 12.10. A wrongly placed conductor can lead to unexpected fault conditions occurring. Make a written note where the conductors are terminated when servicing.

the task in hand can often lead to unnecessary frustration when left with a group of unmarked conductors with little knowledge of how they were terminated. Figure 12.10 illustrates this principle.

As with all test procedures, remember to check the meter both before and after use.

Recapitulation At first sight, wiring a complicated control system for an average size domestic central heating installation can seem quite a daunting task, but it is no more difficult than any other electrical undertaking, except that it may be unfamiliar.

Bear in mind the principal points debated:

1. Provide means of isolation by use of a double pole switch.
2. Wire using a lighting grade cable.
3. Identify the role of each cable.
4. Add cable markers to each conductor.

5. Connect with care.
6. Check constantly to avoid mistakes.
7. Test, using a good quality multimeter.

Finally, work at a pace which suits you best.

The central joint box method

Many central heating systems have been purposely designed to accommodate the central joint box method of wiring. For access and ease of installation, a suitably sized insulated box housing a row of fixed connectors, screwed securely to the rear, will need to be provided. Each controlled circuit is then terminated within the adaptable box where all interconnections are made. Figure 12.11 illustrates a typical example of the form an enclosure might take. Termination techniques follow similarly to the direct wiring method but it is useful to keep the written destination on the outer cable sheath after entering the central joint box. In this way it is easier to make good should fault-finding become necessary.

Wiring from the central joint box A basic block circuit diagram is invariably supplied with the programmer to enable the installation to be carried through with the minimum of trouble as illustrated in Figure 12.12. Each allotted conductor is numbered at either end and is terminated to match the numbered block connectors fixed within the central joint box and the accessory served. Care should be taken when connecting. Always prepare

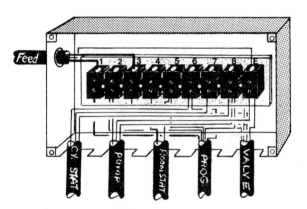

Figure 12.11. Central heating controls; central joint box method.

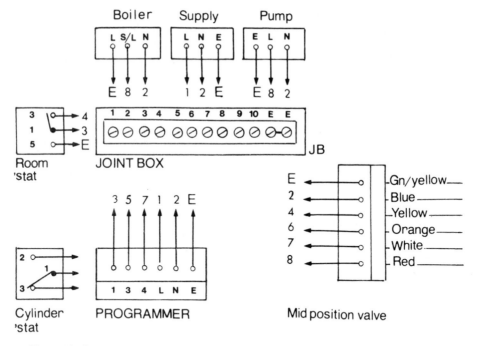

Figure 12.12. Central heating controls: a typical block circuit diagram method of installation.

one cable at a time and attach a numbered identification tag to each conductor, reflecting the terminal it occupies. Leave the destination written on the cable sheath for future reference and try to work in an atmosphere free from distractions. This way, mistakes can be minimised or avoided completely.

The advantage of this method is that it reduces the number of cables entering the programmer for connection. By using a central joint box all conductors may be terminated easily. This will help prevent cable crowding.

All control and current-consuming devices are obtainable as a single package in order to facilitate a smooth and straightforward installation.

Cable terminations High temperatures will cause PVC insulated and sheathed cables to deteriorate, so it is prudent to terminate the fixed wiring installation into a suitable flex outlet accessory before connection is made to the central heating appliance. Final connection may then be carried out by means of heat-resistant multicored flexible cable. This method is ideally suited for cylinder

thermostats, boilers, circulation pumps and mid-position diverter valves, but thermostats, frost thermostats and programmers should always be wired directly.

Many electricians prefer the central joint box method as all interconnections are made within one large adaptable box making it far easier to wire.

Summary

1. Permanent damage can result from careless working practices when an element is removed from a hot water storage vessel.
2. Approach the task in a logical and pragmatic manner. Haste, coupled with insufficient thought, is incompatible with an assignment of this nature.
3. Use the correct tools to meet the demands of the job. Should rewiring be required, use the correct sized heat-resistant cable.
4. Bear in mind the details considered and the following three underlying principles:

   (a) *care* (whilst removing and fitting);
   (b) *clean* (limescale contaminated areas);

(c) *secure* (by means of jointing paste and the correct tools).

Domestic central heating: wiring and controls

Remember the merits of both wiring techniques and keep in mind the following basic considerations:

5. Choose a suitably sized enclosure (insulated).
6. Fit and number a 12-way strip connector block to the rear of the enclosure (nylon or bakelite).
7. Wire the installation throughout using a lighting grade cable.
8. Identify the role played or destination of each cable by marking the outer sheath.
9. Apply numerical cable markers to each conductor corresponding with terminals occupied.
10. Connect central heating appliances by means of heat-resistant multicored cable.
11. Connect with care. Avoid distractions.
12. Check progress frequently.

Handy hints

- Never take for granted that a circuit is 'dead'. Always test using a reliable voltmeter.
- For an emergency continuity tester a 6 V bell or buzzer may be wired to a suitable battery.
- PVC electrical tape can often become hard and brittle during the colder months, but if placed in an overall pocket will absorb sufficient warmth to keep it in a supple and serviceable condition.

13 Practical capacitors and motors

In this chapter: Practical capacitors. Capacitor malfunction and replacement. Starters and means of control. Eddy currents. Overcurrent protection. Thermal overloads. Speed control. DC motors: starting methods. AC single phase motors. Testing. Centrifugal switches. Current relays. General fault conditions. Star—delta starters. Three phase machines and motor maintenance.

Practical capacitors

In the early days of electrical science, capacitors were known as *condensers* and were used extensively in wireless telegraphy. Today they play an ever-increasing role in electrical engineering and are used to improve both torque and phase differential in single phase motors. In the electronics industry they are commonly used in smoothing circuits. With so varied a usage, the capacitor can only be described as both versatile and multifunctional.

Each of the many varieties of capacitors has its own particular role to play. Air capacitors, Figure 13.1, are used exclusively in radio tuning circuits whereas mica and ceramic varieties are generally

employed for high frequency work such as radar and television.

In low frequency electrical engineering two types of capacitors are generally used:

1. *Paper insulated, oil-filled capacitor* — widely accepted for many applications and can vary in value from a few picofarads to many microfarads. Mainly employed in motor and power factor correction circuits. See also Chapter 4.
2. *AC electrolytic capacitor* — designed only to be in circuit for short periods at a time. This type of capacitor is capable of storing large amounts of capacitance in relation to its size, so is ideally suited to being used as a starting capacitor for small induction motors. Many manufacturers recommend a limiting time of between six to eight starts per hour; each start being of no more than 10 seconds in duration. Should an electrolytic capacitor be wired permanently in circuit, severe damage could incur resulting in total breakdown.

Figure 13.2 illustrates the fundamental difference between the paper insulated, oil-filled and electrolytic capacitors.

General construction

The most basic form of capacitance may be demonstrated by placing two plates of metal together but separated from each other by an insulating material called a *dielectric*, as shown in Figure 13.3. The total capacitance, measured in farads or subdivisions of farads, is totally dependent on the surface area of the plates, the distance from each other and the material composition of the dielectric. Obviously, the working voltage must also be taken into consideration: too high, and the dielectric will break down, causing the capacitor to malfunction.

Figure 13.1. Air capacitor used in radio tuning.

Figure 13.2. From left to right: paper insulated oil-filled capacitor; electrolytic capacitor.

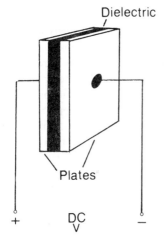

Figure 13.3. A basic capacitor.

The voltage per unit thickness of the dielectric determines the working efficiency of the capacitor, a list of which was given at the beginning of Chapter 5.

In practical terms, the paper insulated, oil-filled capacitor is constructed from a continuous length of oil-impregnated paper, often many metres long. This is fashioned to form the dielectric and is sandwiched between two equivalent lengths of thin foil but of slightly smaller diameter. Connections are made to the tin foil plates by way of tabs and the complete assemblage is rolled and placed into a suitable container.

Capacitor malfunction

Capacitor failure can be contributed to a variety of reasons, many of which could be avoided by prudent maintenance. The more commonplace causes why breakdowns occur under load conditions are as follows:

1. *Working voltage of the capacitor rated lower than the voltage applied* — an increased voltage will give rise to the dielectric breaking down which will inevitably lead to short circuit conditions appearing between the plates.
2. *Faulty starting or switching apparatus* — many manufacturers recommend electrolytic capacitors to be in circuit for a maximum of 10 seconds. Should this time be exceeded, because of electrical failure of the starting apparatus or mechanical malfunction within the automatic switching device, the capacitor will break down.
3. *Excessive temperature* — Pressure accumulating within a capacitor can cause the housing to rupture, leading to an irretrievable loss of oil.
4. *Poor ventilation* — this can contribute to capacitor malfunction owing to a build up of ambient temperature.
5. *Dampness and general contamination* — this condition is often responsible for leakages of current tracking from live conductive parts to the protective metal casing.
6. *Prolonged or frequent starting* — prolonged or frequent starting will cause the oxide insulating film acting as an electrolytic dielectric to break down. If frequent starting is unavoidable, a paper insulated, oil-filled capacitor of the same value, but higher working voltage, is recommended.

A useful practical test

Other than a mathematical evaluation, as examined in Chapter 5, there is little which may be done to test the validity of a capacitor under site conditions. As an alternative to the theoretical approach, a straightforward insulation test can be implemented. Figure 13.4 describes how such a test is carried out. Care should be taken that

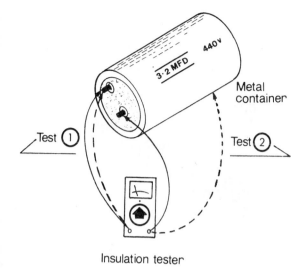

Figure 13.4. Carrying out an on-site insulation test on a capacitor.

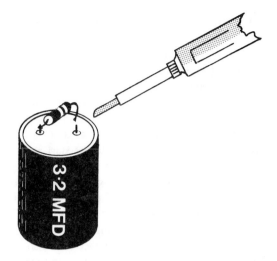

Figure 13.5. Unsolder one side of the discharge resistor before testing is carried out.

voltages in excess of the normal working voltage of the capacitor are not used, as this will prevent dielectric breakdown occurring. When an insulation tester is placed across the terminals of a healthy capacitor, the reading is momentarily deflected towards 'Zero' before creeping slowly towards infinity, whereas a short circuited capacitor will provide a sustained reading of 'Zero' or approaching 'Zero'.

Some capacitors are designed with high value resistors soldered between their terminal poles making it difficult to carry out an insulation test of this nature. To remedy, unsolder one side of the resistor before testing as shown in Figure 13.5. A fault condition existing between either plate and the metal casing will be instrumentally indicated as a low resistance value in ohms. Capacitors may also be charged with an insulation test meter, set to the correct working voltage of the capacitor, providing the charging leads are removed before the test is terminated. A healthy capacitor will produce an audible 'crack' accompanied by an illuminated display when the terminal poles are shorted out. Care should be taken when carrying out this exercise as capacitors are able to store a great deal of energy.

Unfortunately, this is not an accurate test since there is no way in which broken plates may be

identified or to detect whether the dielectric is gradually being broken down by the applied voltage. However, this practical test may prove helpful when in a difficult situation under site conditions. Ideally, suspect capacitors should be tested in a workshop using a *capacitor analyser meter*.

On completion, resolder the discharge resistor in its original position. Loose or stock resistors may also be used to demobilise stock capacitors suspected of being fully charged.

Replacement capacitors: motors

The value of a replacement capacitor should equal the original in order to maintain identical torque and phase differential. If this is not practical, a replacement starting capacitor should be no greater than ± 20 per cent of its original value, or ± 10 per cent should a run winding capacitor be replaced. A value greater than the original will permit current flow through the start winding to be more than 90 degrees out of phase with the run winding. This will affect the performance of the starting torque and so reduce the efficiency of the machine.

It is important to record the original value whenever a replacement differs. This may be done simply by writing the details, using a permanent marker, on the stator housing so others who follow will be aware.

Whenever replacement is necessary, ensure that the working voltage is correct. The correct value, but with an operating voltage of 50 V, would soon break down if incorporated within a 230 V supply! A split phase induction motor will always have in attendance a 400 V class capacitor as it is continuously in circuit throughout the start and run sequence. However, a capacitor start induction motor employing a centrifugal switch, enabling the start winding to be taken out of circuit, need only be served with a 230 V working capacitor. This is because it need only be in circuit for a short duration during the starting sequence (see also Chapters 4 and 5). This type of motor will also run if briskly flicked should a breakdown occur resulting from capacitor malfunction. Care must always be taken when carrying out this hazardous test and if you are at all doubtful, resort to a less experimental method of determining the nature of the fault condition.

Handy hints
- Two fuse elements wired in parallel formation will reduce their joint current-carrying capacity by approximately 8 per cent; three by 15 per cent.
- A good magnifying glass will prove a useful tool for reading small or faded print often accompanying older electrical equipment.
- Figure 13.6 outlines how a remedial grommet may be made by slicing edgeways a suitable length of three core PVC insulated and sheathed cable. By removing the cores and snipping the sliced edge two-thirds across the diameter of the empty sheath, the prepared workpiece may be manipulated to accommodate any size or shaped hole.

Practical motors and means of control

Theoretical concepts examined in Chapter 4 provided the necessary groundwork to enable more practical aspects to be evaluated.

Remove conductors

Cut sheathing

Figure 13.6. A grommet strip can be made from a suitable length of PVC insulated and sheathed cable.

Means of control and operation
Figure 13.7 illustrates diagrammatically a typical wiring arrangement serving a single phase electric motor. In the UK, the *Wiring Regulations* require that a means must be provided to prevent an electric motor from restarting after stoppage due to voltage failure or overcurrent. To satisfy Regulation 522 an automatically controlled electromagnetic starting device should be used, but this stipulation is not applicable to motors under 0.37 kW or if serving a pump controlled by a float switch, as Figure 13.8 depicts. Regulation 552-01-02 confirms.

The starter operates on the principle of converting electrical energy into linear motion by means of an electromagnetic coil in tandem with a mechanical switching assembly. This may be energised either manually or by automatic means such as a float switch or thermostat. Figure 13.9 shows schematic detail of the internal wiring arrangement for a typical three phase direct-on-line, manually operated starter. Once the starting sequence has been brought into play, Terminals 7 and 8 are simultaneously bridged, thus providing a continuous voltage to the coil.

The generated electromagnetic flux provides for linear motion enabling the laminated armature, forming an integral part of the mobile switching mechanism, to overcome the strength of the service springs, and to be drawn magnetically to the iron core. Once energised, the switching assembly automatically maintains a supply to the

TABLE 13.1 Correlation between voltage, current and torque
(courtesy of Salisbury College)

Percentage of declared voltage (U_0)	Starting current (percentage of full load)	Percentage of starting torque
35	76	11
40	110	24
62	260	57
80	384	96

Should the motor fail to respond to the minimum voltage range, output tappings may be adjusted to a higher value. Advantages gained from using this method of speed control include:

1. Reduced starting current.
2. Torque proportional to voltage applied.
3. Smoother starting sequence.

Disadvantages are:

1. High installation costs.
2. Regular maintenance required.
3. Only certain types of motors are suitable.

Table 13.1 highlights the relationship between the supply voltage, current and torque when a typical capacitor aided induction motor is placed in circuit with an auto transformer speed controller. At 35 per cent of full voltage the starting torque is just 11 per cent. This would be quite adequate to serve a small load but starting could prove more of a problem if the load were greater.

Regulation 55-01-02 forbids the use of a step-up auto transformer in an IT supply system, a system not used for public supplies where there is no direct connection between any live part and earth. The consumer's installation is earthed and protected by means of a residual current device.

Speed control: DC machines

DC machines are mechanically started by providing an inverse variable resistance to both armature and field coil in unison.

Figure 13.19 demonstrates how this method is applied to a typical DC shunt motor.

1. *Position 1* shows the starter with maximum

resistance in the armature and no resistance in the field coil circuit.
2. *Position 4* illustrates the control circuit in a mid-position mode; the total resistance shared equally between both armature and field coil.
3. *Position 7* depicts the rotor arm magnetically held to the 'no-volt-coil'. All resistance is mechanically removed from the armature circuit at this stage, transferring maximum resistance to the field coil.

Protection against overcurrent

Since both armature and overload coil are arranged in series formation, an overcurrent condition would produce a proportional increase in current throughout the circuit. This would allow sufficient magnetic flux to be generated to attract the accompanying 'no-volt' short circuiting terminal bar. Once shorted, the accompanying magnetic field would collapse whilst simultaneously releasing the spring loaded rotor arm and effectively isolating the supply to the motor.

DC linear resistance speed control Figure 13.20 illustrates schematically a linear resistance field regulator placed in series formation with a shunt field winding. The strength of the magnetic flux can be altered by varying the field current. An increase in resistance will proportionally decrease the flux and so increase the speed of the armature.

Voltage differential The majority of motors manufactured today are designed to operate efficiently between ± 5 per cent of the declared voltage. This implies that a machine intended for use on a 230 V supply could run satisfactorily between 218.5 and 241.5 V.

Practical single phase machines

Five commonplace single phase motors in general use today are now described.

Capacitor start induction motor

Mechanical characteristics

1. Relay or centrifugal switch employed in

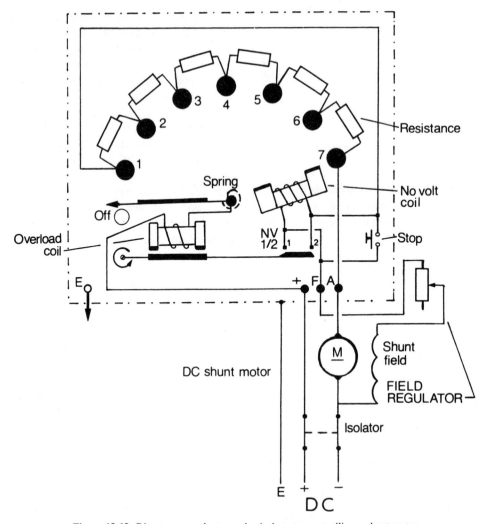

Figure 13.19. Direct current electromechanical starter controlling a shunt motor.

circuit to isolate the start winding when 70 per cent of the rotor speed has been reached.

2. Capacitor fitted to the stator housing or remotely sited.

3. Wound stator, squirrel caged rotor.

Running characteristic Good starting torque but a slight drop in speed with load.

Reversed rotation Interchange leads serving the start winding as shown in Figure 13.21. These are identified as Z1 and Z2.

Practical applications Pumps, compressors and motors above 0.2 kW designed to start against load.

Supplementary information

1. Capacitors used are usually the 230 V working range electrolytic variety.

2. High impedance start winding.

Series motor

Mechanical characteristics

1. Wound armature served by a brass segmented commutator.

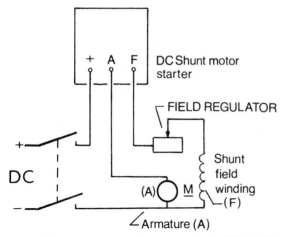

Figure 13.20. Wired in series formation with the field winding, a field regulator will provide for speed control.

2. Current supplied to the armature by means of two spring loaded carbon brushes diametrically placed on the periphery of the commutator.
3. Wound field coils.

Running characteristics

1. Good starting torque but a marked reduction of speed with load.
2. Constant speed difficult to achieve.

Reversed rotation Interchange the internal leads serving the carbon brushes as illustrated in Figure 13.22.

Practical applications

1. Small domestic appliances.
2. Portable hand-held electric drills.
3. Electric tools rated at below 0.2 kW.

Supplementary information Armature and field winding placed in series formation.

Shaded pole induction motor

Mechanical characteristics

1. Wound stator, squirrel cage rotor.
2. Shading poles formed on stator.
3. No start winding, capacitor or centrifugal switch.

Running characteristic Slight drop in speed with load.

Reversed rotation Not very practical; usually classed as non-reversible. However, theoretically, reversed rotation may be attained by removing the stator windings and physically turning them around.

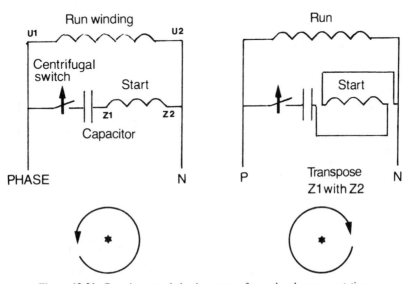

Figure 13.21. Capacitor start induction motor: forward and reverse rotation.

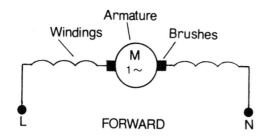
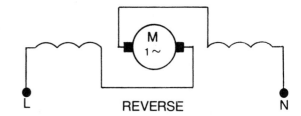

Figure 13.22. Series motor. Forward and reverse rotation.

Practical applications

1. Small refrigeration fans.
2. Desk fans.
3. Record deck drives.

Supplementary information Reliable and virtually trouble-free.

Induction split phase

Mechanical characteristics

1. Relay or centrifugal switch employed to isolate the start winding once approximately 70 per cent of the rotor speed has been reached.
2. Start and run windings wired in parallel formation.
3. Squirrel cage rotor.

Running characteristic Poor starting torque and a slight drop in speed when load is applied.

Reversed rotation Interchange leads serving the start winding. These are illustrated in Figure 13.23 as Z1 and Z2.

Practical applications

1. Used under conditions where the starting load is insignificant.
2. Generally limited to motors over 0.2 kW.
3. Examples include workshop pedestal drills and industrial vacuum cleaners.

Supplementary information

1. Start winding has a higher impedance than the run winding.
2. No capacitor in circuit.

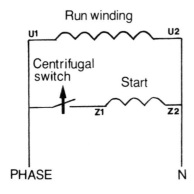

Figure 13.23. Induction split phase motor: forward and reverse rotation.

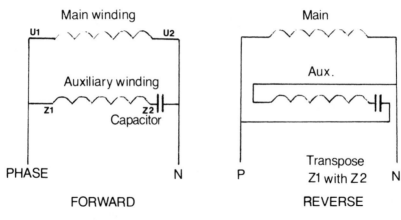

Figure 13.24. Induction split phase motor (permanent capacitor): forward and reverse rotation.

Induction split phase (permanent capacitor)

Mechanical characteristics

1. Capacitor permanently in series formation with auxiliary winding.
2. Run and auxiliary windings wired in parallel formation.
3. Squirrel cage rotor.

Running characteristics Poor starting torque and a slight drop in speed experienced under load conditions.

Reversed rotation Interchange internal wires leading to the auxiliary winding marked Z1 and Z2 as shown in Figure 13.24.

Practical applications Used for light loads or where rapid acceleration is not required, such as unit heaters and refrigeration condenser fans.

Supplementary information

1. Often a second capacitor is placed in parallel with the run capacitor as an aid to starting and is automatically taken out of circuit by means of a centrifugal switch once the rotor has attained approximately 70 per cent of its potential speed (Figure 13.25).
2. Generally oil-filled 400 V working voltage capacitors are used.

3. The auxiliary winding has a higher impedance than the run winding and remains continually in circuit.

Testing for continuity
Continuity testing may be carried out in a reasonably straightforward manner.

Disconnect the two sets of windings at the terminal housing, making a written note of how they are connected. Tails leading to the start and run windings can be identified as Z1 and Z2, U1 and U2, respectively, although older machines are run wind coded A1 and A2; a yesteryear equivalent to U1 and U2.

Ideally, a low scaled digital ohmmeter should be used to measure the impedance offered across the

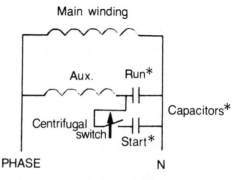

Figure 13.25. To aid starting, an auxiliary capacitor is placed in parallel formation with the run capacitor and taken out of circuit automatically.

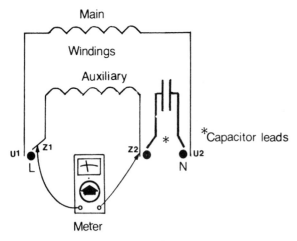

Figure 13.26. Continuity testing a split phase machine (see also Fig. 13.24).

individual sets of field windings as Figure 13.26 shows. The start winding will always record a higher resistance value than the run and, depending on the design of the motor, could be up to 10 times greater in value.

This is an ideal test to prove the continuity of the field coils. It will not indicate whether any part of the winding has been short circuited because of mechanical damage or overcurrent. To establish this, knowledge must first be gained as to the total impedance which may be envisaged for both sets of windings.

The centrifugal switch

The role of the centrifugal switch is to disconnect the start winding from the supply once the rotor has reached between 70 and 80 per cent of its potential speed. Many are disc-like, accompanied by a duo of counterweights and control springs.

Overcome by the centrifugal movement of the counterweights, the control springs yield, enabling the switching disc to function. Once opened the start winding is effectively disconnected from the supply. Figure 13.27 illustrates this mechanical principle in simple diagrammatical form. When the machine is electrically isolated the reverse happens and the switching disc is snapped back to its original position. This effectively closes the contacts as the motor loses speed and is primed for the next starting sequence.

Problems resulting from centrifugal switching devices

Practical difficulties may appear in many different ways, frequently resulting in the motor not being able to start or to change from one running mode to another. Listed are three common conditions which can affect the automatic switching sequence.

1. *Switching action permanently made.* This often results from broken springs or arc welded switching contacts. To rectify, change the switching assembly for a correct replacement. Check and, if necessary, replace the capacitor. Check and test both sets of windings for evidence of overcurrent.
2. *Switching mode permanently open.* Generally resulting from a mechanical breakdown within the centrifugal switch. Machines falling casualty to this condition will develop a '50 cycle hum' which will persist until closed down by means of local overcurrent protection. The run winding should be checked for evidence of overcurrent.
3. *Switching assembly burnt out.* Usually caused by excess current developing in the start winding or resulting from a faulty capacitor. Check both.

The current relay

A current operated relay is used as an alternative to the centrifugal switch where such a device might prove impractical when, for example, serving a submersible slurry or water pump.

Made from a few turns of heavy gauge, but lightly insulated, wire, the *relay coil* is remotely placed in *series* formation with the run windings as illustrated in Figure 13.28. On starting, both run winding and relay coil carry a high current, producing a strong magnetic flux throughout the coil. Once sufficiently energised the relay switching assembly is activated, bringing the start winding into play. When between 70 and 80 per cent of full load speed has been reached, current flowing through the run winding diminishes. This proportionately affects current flowing through the relay coil since both are wired in series formation. The magnetic flux then subsides sufficiently to

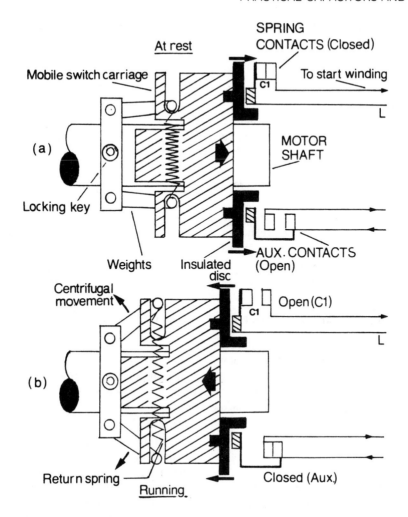

Figure 13.27. The centrifugal switch: (a) at rest; (b) running.

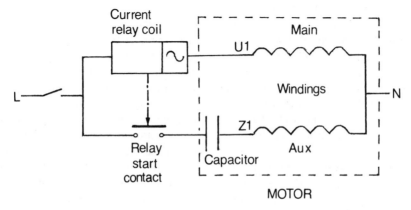

Figure 13.28. The current relay.

allow the start winding contacts to reopen under gravity.

Testing

To test, physically check the mechanical action of the relay's switching assembly, making sure that it opens smoothly. Inspect the contact points and clean if required. A continuity test should then be carried out across the coil and between the switching contacts of the relay. Check for loose terminals as these can often contribute to future fault conditions if left unattended.

Replacement relays must always be fitted in the correct position so that the de-energised relay can open under gravity. Some units are factory sealed, making it necessary to provide a complete replacement should a fault condition occur.

General fault conditions: single phase machines

Although reliable and virtually trouble-free during the majority of its working life, the electric motor can occasionally develop problems and fault conditions which call for professional attention.

Listed are eight of the more common weaknesses which can occur during the lifetime of the machine.

Undulating fan motors

This is often caused by the fan blade assembly slipping on the motor shaft or variations in the supply voltage.

Rust-seized fan assembly

A simple but effective method of removal when pulley extractors are unavailable is to play a *gentle* flame around the neck of the fan blades, Figure 13.29. Rotating the blades will ensure that heat is applied evenly but care must be taken to avoid possible damage to the bearings. The expanded fan can be removed by gently tapping a copper drift with a suitable hammer around the neck of the fan once machine screws, used to secure the the fan assembly to the motor shaft, have been loosened. This technique should not be used for plastic fans.

When replacing, apply a smear of petroleum jelly on the leading face of the motor shaft. This

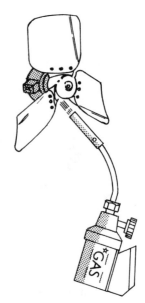

Figure 13.29. Play a gentle flame around the neck of the fan blade assembly.

will help prevent rust from building up, so reducing future problems.

Overheating

A small domestic appliance motor found to be running hot, yet drawing a nominal current, is often a victim of dry soldered joints or badly made internal compression connections. The machine will vary in both speed and audibility and will maintain a higher than normal temperature. To remedy, resolder the internal joints and check any compression connectors.

Worn bearings

Electric motor bearings will increase in wear when pulley belts are badly aligned. If left, damage from continuous use will result. Both load and motor drives may be easily aligned using a long metal 'straight edge' as illustrated in Figure 13.30. It is advisable to soften the belts before coupling and allow approximately 12 mm of play after fitting. Additional play would probably lead to premature wear whilst on load.

Under- or over-greased bearings are also a contributory cause of premature wear.

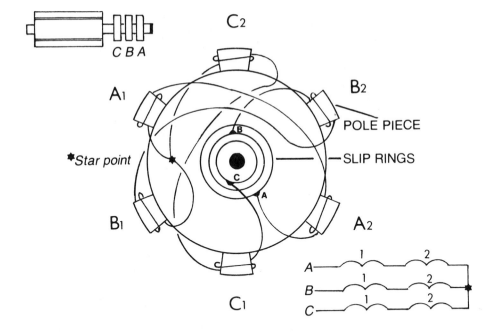

Figure 13.36. The rotor of a wound induction motor — an illustrative concept. (Because of its uniqueness, this motor has been annotated differently using capital letters (unlike Figure 13.33). A1–A2: B1–B2: C1–C2 are pairs of *rotor* poles made common at the slip ring CBA. These are graphically shown in the top left corner in side view. Figure 13.36, although similar to Figure 13.33, depicts an electric motor's ROTOR windings (the moving part). ABC at the bottom right corner depicts the end of the rotor windings connected to the slip rings, shown in the top left hand corner.

Figure 13.37. Stator: induction motor.

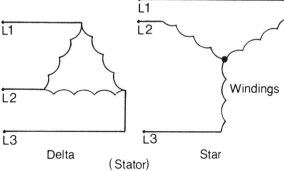

Figure 13.38. Triple phase motor windings drawn schematically in both star and delta formations.

Testing

1. A low scaled ohmmeter should be used to check the continuity of the field windings.
2. Difficult to establish short circuit conditions.
3. Check with an insulation tester between each machine winding and the motor enclosure for possible earth faults.

3. Heavy starting current (up to 600 per cent of full load but this can be reduced to 400 per cent when a double caged induction motor is used). Regulation 525-01-02 confirms.

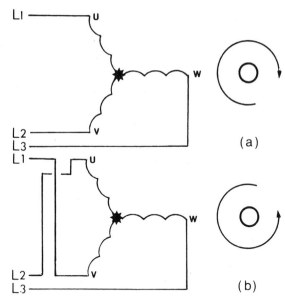

L1 —— U
W
L2 —— V
L3 ——
(a)

L1 ——
U
W
L2 ——
V
L3 ——
(b)

Figure 13.39. Three phase induction motor: forward and reverse rotation (U, V and W are integral motor labels indicating the three internal windings).

Wound rotor induction motor

Mechanical characteristics

1. Wound rotor and stator, wired in star formation.
2. Rotor windings terminated to three slip rings mounted on the shaft. These are mechanically interlinked by way of carbon or soft copper brushes serving a remotely located resistance bank as illustrated in Figure 13.40.

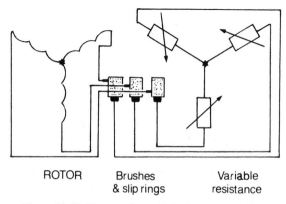

ROTOR   Brushes   Variable
               & slip rings   resistance

Figure 13.40. The wound rotor induction motor: method of speed control (stator not included).

3. Current flow in the rotor may be regulated by means of an external control circuit
4. Both rotor and stator windings are equal in numbers.

Running characteristics

1. High starting torque.
2. With all external resistance removed, the rotor acts as a typical squirrel cage motor.

Reversed rotation Interchange any two supply phases as detailed in Figure 13.41.

Practical applications

1. Large air conditioning plants.
2. Large air compressors.

Supplementary information

1. Stator connected to the supply.
2. Rotor connected to a variable external resistance bank.
3. Low starting current.

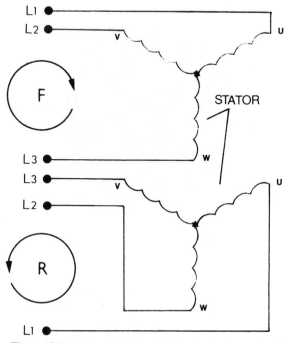

Figure 13.41. The wound rotor induction motor: forward and reverse rotation (rotor not included).

4. Starting by means of a *current limit control*, a method by which control sensors are placed in circuit to detect current flow in the stator.

Testing

1. Rotor winding continuity may be confirmed by placing the test leads of a low scaled ohmmeter between the slip rings. Each combination must tender a similar reading as explained in Figure 13.42. It is often difficult to test for short circuits under site or emergency conditions. Special instruments designed to measure the inductance of the windings are used where there is doubt as to the precise nature of the fault condition.
2. The stator may be tested for open circuit in a similar manner.
3. A further test should be carried out between each slip ring and the shaft of the motor to check for potential earth fault conditions.

This is best achieved by use of an insulation test meter.

The synchronous motor

In Chapter 4 consideration was given to a simple synchronous motor in which the rotor was formed from a strong permanent magnet. This type of machine is used exclusively for light loads, and because of its renowned reliability and constant speed, it is frequently incorporated into the design of electric clocks.

Larger synchronous motors are better developed and are far more sophisticated than their smaller single phase cousins. At first sight the motor appears to be a hybrid; a mixture of both induction and wound rotor induction as Figure 13.43 shows. The rotor is equipped with squirrel cage bars constructed in a similar fashion to a typical induction rotor. These circumnavigate the outer periphery of the rotor. High resistance coils are placed in series formation below the squirrel cage and are served by two slip rings mounted on the shaft.

A three phase supply, connected to the stator windings, produces a rotating magnetic field and allows current to be induced into the squirrel cage each time the stator's magnetic field cuts through the bars. Since the cage is short circuited, current will flow continuously, creating its own independent magnetic field and interacting with that of the stators. This has the effect of dragging the rotor around, trying to pace the stator's gyrating magnetic field.

When sufficient momentum has been gained to just below the synchronous speed, a DC supply is

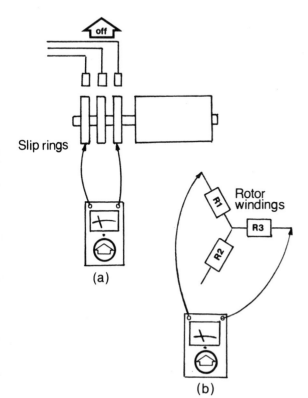

Figure 13.42. The wound rotor: a simple continuity test.

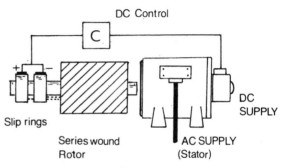

Figure 13.43. The synchronous motor.

automatically switched to provide current to the series wound rotor by way of the slip rings. As the rotor approaches synchronous speed the magnetic cutting action through the bars of the squirrel cage lessens and the current is greatly reduced until finally diminishing when the synchronous speed has been reached.

The energised series wound rotor produces an electromagnetic field which locks on to the stator's rotating field. At this point the rotor will spin in unison with the gyrating stator field.

Since the magnetic field is cutting through both squirrel cage *and* rotor windings at high speed, considerable amounts of current are inadvertently induced in the DC rotor windings when the stator is first energised. To reduce this unwanted current, a field discharge resistor (Figure 13.44) is wired across the series windings. This has an additional effect of preventing induced currents entering the squirrel cage rotor when the magnetic field collapses once the DC supply has been isolated.

Direct current is normally provided by a small generator mounted on the shaft of the rotor as shown in Figure 13.43.

Mechanical characteristics

1. Wound rotor, wired in series formation.
2. Squirrel cage circumnavigating the outer periphery of the rotor.

3. Series rotor windings served by two slip rings mounted on the shaft.
4. Field discharge resistor employed to reduce unwanted induced currents within the rotor.

Running characteristics

1. High starting torque but will stall if the load is too great.
2. Low starting current.
3. Will run in constant unison with the synchronous speed.

Reversed rotation Interchange any two supply phases to the stator circuit (Figure 13.45).

Practical applications

1. Large air conditioning plants.
2. Situations where large motors are required to run at a constant speed.
3. As a means of power factor correction.

Supplementary information

1. Series rotor windings supplied from a small DC generator attached to the shaft of the machine or by other independent means.
2. Supply to the rotor is automatically switched

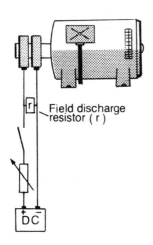

Figure 13.44. To reduce unwanted current, a field discharge resistor is wired across the series rotor winding.

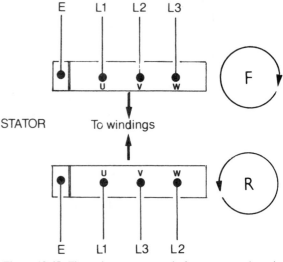

Figure 13.45. Three phase motor terminal arrangement (stator) for a synchronous motor showing both forward (F) and reverse (R) rotation.

in circuit when approximately 95 per cent of the synchronous speed has been reached.

3. High costs.

Testing

1. A low scaled ohmmeter can be used to check the rotor windings for continuity by placing the test probes across the slip rings as described in Figure 13.46.
2. By checking the electrical specifications of the DC generator and applying Ohm's Law, the total resistance of the rotor coils may be calculated.
3. Check the stator windings between each phase, again by use of a low scaled ohmmeter. Each combination must offer similar readings.
4. A DC voltage output test will confirm the health of the generator.
5. An insulation test should be carried out between each phase and the motor enclosure and between each slip ring and the shaft to

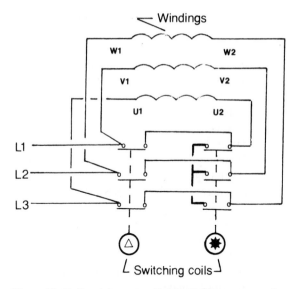

Figure 13.47. Star-delta starter. Basic switching arrangement. (O/L=overload (an automatic switch).)

check for leakages to earth. A high or infinitive test value will indicate an earth fault-free machine.

The star–delta starter

Figure 13.47 illustrates the basic wiring and switching arrangements serving a typical star–delta starter, whilst Figure 13.48 offers a step by step analysis of the control circuit switching sequence.

Once the 'start' button has been manually activated automatic switching gains control in the following sequence:

1. Coil Y is energised bringing in switching contact Y1.

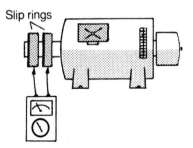

Figure 13.46. Use a low scaled ohmmeter both to measure and to check continuity of the rotor windings.

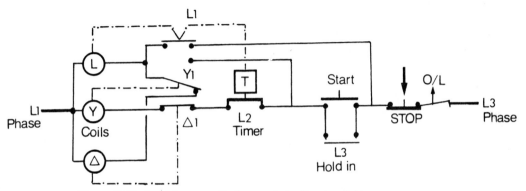

Figure 13.48. Star-delta starter. Basic control circuit and switching sequence.

2. The motor windings are now in star formation.
3. The two way switching contact also provides a path to energise coil L. (For practical purposes coils Y and L are energised instantaneously.)
4. With coil L now energised, contact L1 closes.
5. Timer T serving L2 is energised.
6. Switching contact L2 opens at the end of the timing cycle.
7. De-energised coil Y allows two way switching contact Y1 to open.
8. Switching contact Y1 now provides a supply, by means of closed contact L1, to coil Δ.
9. With coil Δ energised, the motor is placed in delta formation.
10. Energising coil Δ opens switching contact Δ1 providing interlocking facilities by preventing coil Y from being energised.
11. By manually activating the 'stop' button, coils Δ and L are de-energised, effectively isolating the supply to the machine.

Motor maintenance

Regular planned maintenance can often prevent expensive and inconvenient breakdowns. Listed, in random order, are 20 aspects which should be considered whenever such work is carried out.

1. Unblock all ventilation grills to allow free circulation of air.
2. Check pulley wheel and drive belt alignment serving both motor and load. Re-align if necessary.
3. Remove contamination and moisture.
4. Examine carbon brushes and renew if past their useful life.
5. Inspect the physical condition of the commutator. Clean if required using a fine grade of abrasive paper.
6. Examine the bearings for wear. Regrease after every10 000 hours.
7. If geared, verify that the oil level is correct. Change at regular intervals.
8. Remove bearings and wash thoroughly with a degreasing agent after two years of continuous running. Inspect for wear. If worn, replace, If satisfactory, repack with suitable grease.
9. Overgreasing can cause premature wear through overheating.
10. Housing caps should be filled with just two-thirds of grease. This will ensure that a supply will always be in contact with the bearing between maintenance periods.
11. Keep all lubricants free from dirt and moisture. Contaminated lubricants will lead to premature bearing deterioration.
12. Breather holes located near the base of a motor should be cleared of contamination.
13. Check the mechanical action of the centrifugal switch by removing the cowl and fixing bolts at the non-drive end of the motor. Clean and remove any impurities. Terminals should be rubbed with a fine abrasive paper to remove any pitting which might have occurred. Confirm electrical continuity of the switch and reassemble.
14. Inspect the condition and flexibility of the belts. Allow approximately a 12 mm displacement as detailed in Figure 13.49.
15. Should only one belt be used in a two pulley system (Figure 13.50), place the working belt nearest to the bearing. Disregarding this rule can cause unnecessary leverage, sufficient to wear the bearings over a period of time.
16. Inspect the condition of the internal ventilation fan. Replace if damaged.
17. Ensure that all cables are secure within the

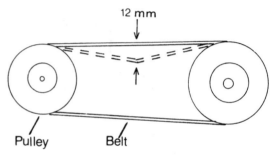

Figure 13.49. Allow a 12 mm displacement when checking the flexibility of drive belts.

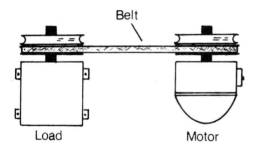

Figure 13.50. A single working belt placed nearside to both load and motor will avoid unnecessary leverage on the bearings.

terminal housing. Check that they are in a satisfactory condition.

18. Carry out a standard insulation test. A damp or contaminated machine provides ideal conditions for faults to prevail.

19. Verify and inspect the condition of the capacitor as described in previous paragraphs.

20. Monitor both starting and running currents using a suitable clamp meter. Remember that the starting current can be up to six times greater than the normal running current on full load.

Summary

1. Capacitors are used to assist power factor correction, to improve torque and phase differential in single phase motors and may also be found in DC smoothing circuits.

2. Oil-filled, paper insulated and electrolytic capacitors are used in low frequency electrical work. Their values range from a few picofarads to many microfarads.

3. The area of the plates, the distance between the plates and the material used to form the dielectric are the principal factors determining capacitance.

4. Capacitor malfunction may be contributed to by:

   (a) working voltage too high;
   (b) faulty starting apparatus, causing prolonged starting;
   (c) excessive temperature;

   (d) poor ventilation;
   (e) mechanical damage;
   (f) dampness and contamination;
   (g) frequent starting.

5. Testing may be carried out using familiar equipment, though it is far better to employ a capacitor analyser meter for reliable evaluation.

6. Replacement capacitors serving an electric motor circuit should ideally be the same value as the original but may vary:

   (a) ±10 per cent ('run' winding);
   (b) ±20 per cent ('start' winding).

7. The working voltage of a replacement capacitor *must* be identical to the original.

Written as an introductory guide, Chapter 13 provides a basic corner-stone on which to build, thus leading to a greater practical understanding of the task to be undertaken.

Considerations were given to both single and three phase machines and means by which control may be affected. Problems involving integral automatic switching, designed to isolate the supply from the start windings once the motor had gained sufficient speed were examined.

Motor overcurrent protection was studied and it was shown that the thermal overload relay was generally accepted as the most reliable means of disconnection.

Speed control for both AC and DC machines was reviewed and found that it was usually only applied for specific purposes in specialised installations.

Review data were offered examining several types of both single and three phase machines.

General fault conditions and maintenance were considered in order to provide valuable insight into the care and conservation of electric motors.

Handy hints

• A pulsating magnetic 'growl' coming from a three phase motor often indicates a missing supply phase. This condition is known as *single phasing*. Check overcurrent protection.

- A loose terminal is sufficient to cause a motor to single phase and activate local overcurrent protection.
- 'Chattering' starter switching assemblies will, if left unattended, lead to the accompanying coil burning out. An over-sensitive probe or temperature control, working with a small differential, is often to blame.

14 Methods of testing

In this chapter: Visual inspections. Testing by instrumentation: protective conductors: continuity of a final ring circuit; insulation resistance; insulation of site-built assemblies; protection by means of electrical separation, by barriers and enclosures. Insulation of non-conducting floors and walls. Polarity and earth fault impedance tests. Earth electrode resistance testing and testing the operation of a residual current device. Periodic inspections.

A precise and detailed examination is both necessary and essential once an installation has been completed. It is important that this should be carried out without reservation. Formal verification of safety must be provided so that hazardous or dangerous conditions which might result in death or injury to an unsuspecting operative are not overlooked.

The Institution of Electrical Engineers has laid down guidelines, within the 16th edition of the *Wiring Regulations*, for both visual and instrumental inspections. These will be reviewed within the next few pages of this chapter. Testing must be strictly observed and thoroughly carried out so that damage to persons, equipment or property is completely avoided.

Testing and procedural recommendations have been drawn from Chapter 71 of the *Wiring Regulations*.

Visual inspections

The visual inspection (from Regulation 712-01-03) may, for practical purposes, be divided into two parts:

1. British Standards inspection.
2. Principal visual inspection.

British Standards inspection (Regulation 511-01-01)
Check first to verify all electrical equipment installed complies with the appropriate British Standard, or is equivalent to that standard, is not damaged to cause a potential hazard and that the installation complies with specifications as laid down within the *Wiring Regulations*.

Principal visual inspection (Regulation 712-01-03)
Listed are 19 suggested items for visual inspection which, if relevant to the installation, need to be inspected before any instrumentation tests are carried out.

Connections Ensure that all connections made to conductors are mechanically sound (Regulation 526-01-01). Ensure good conductance; correct and suitable insulation. Check that protection has been afforded against vibration and mechanical damage.

Ensure all conductors are readily identified This may be carried out by the use of coloured cable, tags, discs and sleeves or by numbering. For conduit installations a distinguishing orange band encompassing the conduits should be applied. BS 1710: 1984 (1989) and Regulation 514-02-01 refer.

Routeing of cables Check the routeing of cables relevant to external and extraneous influences (Regulations 512-06 and 522). This will include cables routed and located in:

1. Airing cupboards.
2. Bathrooms and kitchens.
3. Agricultural environments.
4. Caravans.
5. Boiler houses.

Also the effect of:

1. Ionisation.

2. Ambient temperature.
3. Wind and storm.

This also includes protective measures offered to outside installations.

Conductors Check a random assortment of conductors for both voltage drop and current-carrying capacities.

Cable sizes must be checked to ensure that the current-carrying capacity is not less than the current designed to flow through them. A check should also be made to verify that voltage drop will not exceed 4 per cent of the nominal voltage. Reference is made to Regulation 525-01-02.

The polarities of all switching devices A check should be made on all single pole switching devices to ensure that mechanical isolation occurs in the phase conductor only. Regulation 130-05-01 refers.

The polarity of all socket outlets and lamp-holders The switched phase conductor must only be connected to the middle terminal of an Edison type lamp-holder or ranged to the right, when viewed from the front, when connection is made to a socket outlet (Regulation 713-09-01).

The means of protection against thermal effects and the presence of fire barriers This includes protection offered between floors and walls against thermal effects from over-heating and arcing. Regulation 527-02 refers. Consideration must be given to the positioning of fixed luminaires or lamps and to the installation of distribution boards constructed without a back.

Methods of protection against direct contact This requirement may be met by use of the following measures:

1. By insulation of live conductive parts.
2. By the use of barriers or enclosures.
3. By obstacles.
4. By placing out of reach. As an example: overhead lines.

Reference is made to Regulations 412-03 to 06.

Methods of protection against indirect contact This requirement may be met by one or more of the following measures (Regulation 413-01-01 refers):

1. Automatic disconnection of the supply.
2. Supplementary equipotential bonding (Regulations 413-09 and 471-08).
3. Protection by Class II equipment (Regulation 472-09).
4. Main equipotential bonding.
5. By the presence of protective conductors.
6. Earthing arrangements.
7. Protection by non-conducting location (Regulations 413-05 and 471-11).
8. Earthing conductors.
9. Protection through electrical separation (not to exceed 500 V; Regulations 413-06 and 471-12 refer).
10. Earth-free local equipotential bonding.

Prevention of mutual detrimental influence Equipment to be used should be both selected and erected so as to avoid any potentially damaging or harmful influence to a non-electrical installation in the vicinity. When this proves impractical, both systems must be segregated (Regulation 515-01-01 refers).

All isolating and switching arrangements This will include breakers, fault current protective devices; emergency switching and monitoring equipment. Confirm that they are of the appropriate type and correctly rated for the circuits they control. Inspect to ensure that switching arrangements and devices are unable to reclose through mechanical vibration or shock and are suitably positioned. Regulation 537 refers.

Presence of undervoltage protective devices This requirement may be met by one or more of the following examples in order to monitor equipment liable to become damaged by a drop in voltage.

1. A time delay mechanism may be incorporated.
2. Should the risk be foreseen, then notification must be given to a person responsible for the maintenance of the installation that damage could take place in the event of a drop in voltage.

Chapter 45 and Regulations 451-01-01 to 451-01-06 refer.

Choice and setting of protective monitoring devices A 30 mA tripping rate residual current device is recommended for circuits containing socket outlets. In a TN supply system a maximum disconnection time of 0.4 seconds is called for when the nominal voltage is between 220 and 277 V AC. Regulations 471-08 and 473-01 refer.

Labelling of circuits, switches, fuses and terminals It is recommended that labelling be carried out on the following (Regulation 514-13-01):

1. Earthing arrangements.
2. Equipotential and supplementary bonding.
3. Distribution centres.
4. Switches and isolators.
5. Central joint box terminal arrangements.
6. Buried cables.
7. Position of caravan inlets.
8. Fuses and circuit breakers.
9. Emergency and maintenance switching devices.

Selection of equipment and protective measures appropriate to external influences The term 'external influences' is a measure of one or more of the following:

1. Ambient temperature (Regulation 522-01).
2. External heat sources (Regulation 522-02).
3. Presence of water or high humidity (Regulation 522-03).
4. Presence of solid foreign bodies (Regulation 522-04).
5. Presence of corrosive or polluting substances (Regulation 522-05).
6. Impact (Regulation 522-06).
7. Vibration (Regulation 522-07).
8. Other mechanical stresses (Regulation 522-08).
9. Presence of flora or mould growth (Regulation 522-09).
10. Presence of fauna (animals) (Regulation 522-10).
11. Solar radiation (Regulation 522-11).
12. Building design (Regulation 522-12).

Both type and location of cable and accessories must be considered. Domestic switching arrangements would, for example, be hardly suitable for use out of doors or in an industrial environment.

Suitable access to switchgear and equipment (Regulation 526-04-01 also refers)

1. Routine inspection.
2. Maintenance work.
3. Testing.

Regulation 526-04-01 refers. Other examples may be drawn from the following reasons:

1. Access to cable joints (Regulation 513-01-01).
2. Access to earthing and bonding point connections (Regulation 543-03-03).
3. Access to emergency switching (for example, fireman's switch).

Danger notices Check for the presence of warning or danger notices on or adjacent to electrical equipment. Regulation 514 refers. This will normally incorporate all accessories and apparatus containing voltages in excess of 250 V or where control circuits are supplied by another phase to the equipment served. Typical notices are illustrated as Figure 14.1.

Diagrams Check the availability and presence of diagrams, charts, tables and similar information relevant to the installation. Regulation 514-09-01 refers. This can include one or more of the following:

1. Voltage warning notices.
2. Details of circuit.
3. Method of wiring and size and type of conductor.
4. Destination of circuit.
5. Protective devices installed.
6. Earthing arrangements (for example PME).
7. Periodic inspection notice.
8. Number of points.
9. Notices (for example for touring caravans etc.).

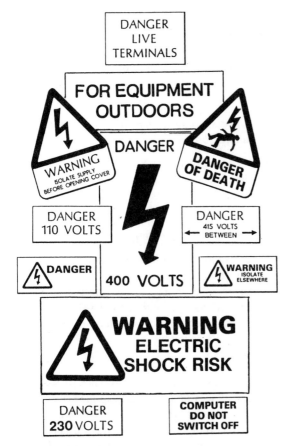

DANGER
LIVE
TERMINALS

FOR EQUIPMENT
OUTDOORS

DANGER

WARNING
ISOLATE SUPPLY
BEFORE OPENING COVER

DANGER
OF DEATH

DANGER
110 VOLTS

DANGER
415 VOLTS
← BETWEEN →

DANGER

400 VOLTS

WARNING
ISOLATE
ELSEWHERE

WARNING
ELECTRIC
SHOCK RISK

DANGER
230 VOLTS

COMPUTER
DO NOT
SWITCH OFF

Figure 14.1. Electrical warning and danger notices. (Reproduced by kind permission of Systems and Electrical Supplies Ltd.)

THIS INSTALLATION, OR PART OF IT, IS PROTECTED BY A DEVICE WHICH AUTOMATICALLY SWITCHES OFF THE SUPPLY IF AN EARTH FAULT DEVELOPS.

TEST QUARTERLY BY PRESSING THE BUTTON MARKED "T" OR "TEST".

THE DEVICE SHOULD SWITCH OFF THE SUPPLY AND SHOULD THEN BE SWITCHED ON TO RESTORE THE SUPPLY.

IF THE DEVICE DOES NOT SWITCH OFF THE SUPPLY WHEN THE BUTTON IS PRESSED, INFORM YOUR ELECTRICAL CONTRACTOR.

Figure 14.2. Advisory notice accompanying an installation served by a residual current or voltage operated circuit breaker. (Reproduced by kind permission of Systems and Electrical Supplies Ltd.)

1. Nominal voltage (U_0)
2. Design current (I_B).
3. Frequency.
4. Power characteristics.
5. Compatibility with other equipment.
6. Conditions likely to be encountered.
7. Accessibility for inspection and maintenance.

Regulations 510-01-01 to 513-01-01 and 120-04 refer.

Practical and instrumental testing

Reference is made to Regulation 713-01-01 to 12. Once visual checks have been completed and proved satisfactory, the following practical and instrumental tests must be carried out in the order stipulated:

1. Continuity of protective and equpotential bonding conductors (Regulation 713-02).
2. Continuity of the final ring circuit (Regulation 713-03).
3. Insulation resistance test (Regulation 713-04)
4. Insulation of site-built assembies (Regulation 713-05).
5. Protection by means of electrical separation (Regulations 713-06 and 413-06-01 to 05).
6. Protection by barriers or enclosures provided during the building of an installation (Regulation 713-07).

10. Fault-operated protective device advisory notice.

An automatic device, designed to interrupt the supply, must always be accompanied by a legible advisory notice. The wording, as recommended in the *Wiring Regulations*, is reproduced unabridged as Figure 14.2.

Erection of equipment Equipment erected must comply with relevant requirements and British and E. U. Standards. Should foreign equipment be used then a check must be carried out to confirm that the equipment used is suitable for one or more of the following examples: .

7. Insulation of non-conducting floors and walls (Regulation 713-08).
8. Polarity testing (Regulation 713-09).
9. Earth fault impedance test (Regulation 713-10).
10. Earth electrode resistance (Regulation 713-11).
11. Testing the operation of a residual current device (Regulation 713-12).

If required, tests 1, 7 and 10 may be carried out before the installation is energised. Regulation 713-11 may be waived should such knowledge be considered unnecessary.

Continuity of protective and equipotential bonding conductors

In order to verify that all circuit protective conductors are electrically sound and securely connected, each conductor, whether forming part of a multicore cable or wired independently, must be individually checked and tested. This will also incorporate cabling used for main and supplementary bonding and any extraneous conductive parts equipotentially bonded.

At first sight this may appear quite a daunting task. There are ways in which the test may be made both simple and straightforward and these will be reviewed below.

Testing circuit protective conductors comprising steel conduit and trunking This test is implemented using a maximum voltage of 50 V AC or DC at one and a half the design current of the circuit to a maximum input of 25 A. Whenever an AC supply is used it must be at the same frequency envisaged for the installation under test. Should a DC voltage be chosen, the protective conductor must be checked throughout its length to ensure that no inductor is incorporated within the circuit.

There are many proprietary test meters available which are ideal to meet the requirements of this regulation. These are known as *conduit continuity and impedance testers*. They are both simple and easy to use and may be plugged into a suitable power source and selected for the frequency required. Two heavy test leads forming an integral part of the meter are clamped to the circuit under test. By means of a manually operated test button, 24 V can be injected into the system (Figure 14.3) and the impedance in ohms read directly from an analogue or digital display meter.

Figure 14.4 shows a proprietary conduit continuity and impedance test meter, whereas

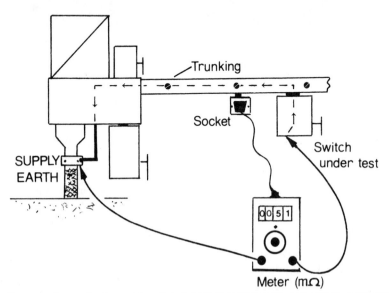

Figure 14.3. Testing protective conductive metal parts and conductors comprising steel conduit and trunking using a conduit continuity and impedance tester.

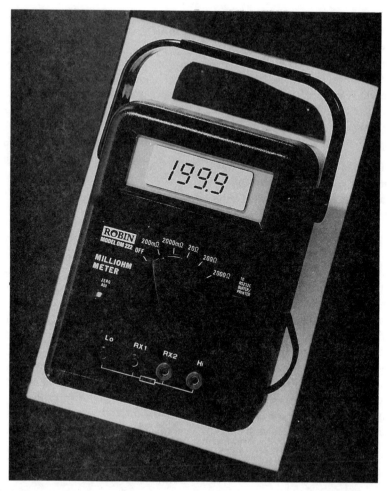

Figure 14.4. Conduit continuity and impedance tester. (*Reproduced by kind permission of Robin Electronics Ltd.*)

Figure 14.5 illustrates schematically the basic wiring arrangements should such an instrument be assembled. Conduit and trunking continuity impedance testing should be regarded as a *secondary test*, used only when a problem has been made apparent through use of an *earth-loop-tester*. Because of the high currents involved, impedance testing should be carried out with extreme care.

Testing protective conductors comprising a single, or forming part of a multicored, cable When a current protective conductor takes the form of a single insulated earth wire or is constructed forming the central member of an insulated and sheathed multicore cable and does not comprise a steel trunking or conduit system, all demands concerning current tests are relinquished. Instead, a standard continuity meter may be used.

To test, attach a long flexible lead to the 'common' service terminal of a continuity meter. The second and accompanying lead need only be short. Connect the free end of the longest lead firmly into the earthing terminal block within the distribution board. This will allow the meter freedom of movement to check all current protective conductors within the installation as Figure 14.6 illustrates. In practice this requires each switch, ceiling point, socket outlet and electrical accessory, including all main equipotential and supplementary bonding to be tested.

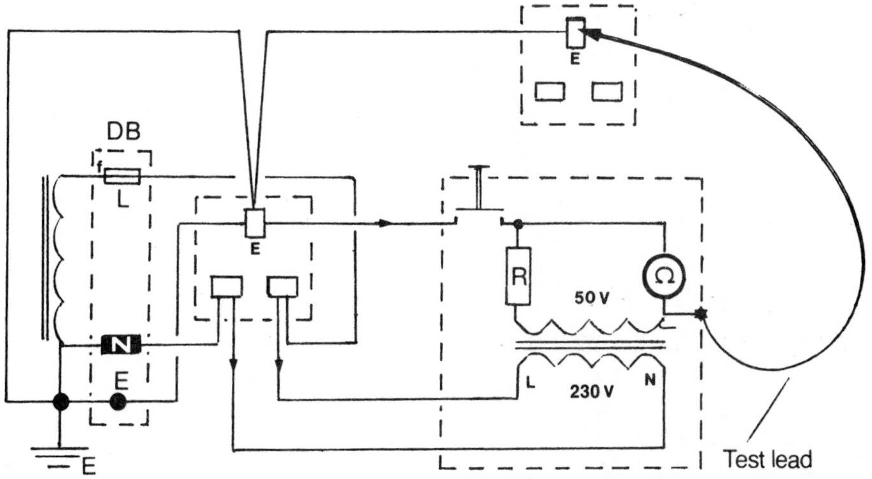

SOCKET UNDER TEST

Socket serving
tester

Transformer
serving loop tester

Test lead

Figure 14.5. Basic integral wiring details: conduit continuity and impedance tester.

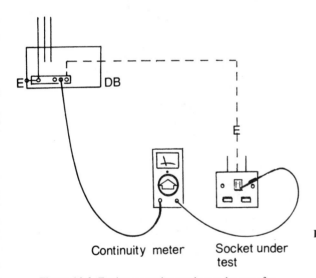

Continuity meter   Socket under
test

Figure 14.6. Testing protective conductors by use of a
continuity meter.

Figure 14.7. Measure the total resistance offered to both long
and short test leads. Deduct this value from future test
readings.

Before testing, measure the total resistance of
the combined flexible leads (Figure 14.7) and
deduct the value obtained from future readings.

Since the internal resistance of a conductor is
directly proportional to length but inversely
proportional to cross-sectional area, it stands to
reason that the further the tests are made from the
distribution centre, the greater the resistance value
of the protective cable will be. Higher than
expected continuity readings should be treated with
suspicion and the cause investigated at once. Often
this is simply due to loose or contaminated
connections or conductors having been overlooked.

Continuity of the final ring circuit
The following procedure can be adopted to meet
the requirements of this regulation. It is
recommended that a low scaled digital continuity

Figure 14.8. Low scale digital continuity meter.

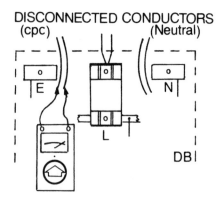

Figure 14.9. Continuity of a final ring circuit. Testing the continuity of the current protective (cpc) and neutral conductors.

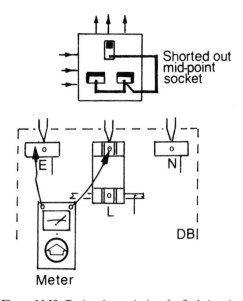

Figure 14.10. Testing the continuity of a final ring circuit. With the mid-point socket shorted out, measure the resistance offered between the phase and current protective conductor at the distribution board.

or multimeter is used in order to carry out this test, illustrated in Figure 14.8.

1. First measure the total resistance offered between the two open ends of the unconnected phase conductor at the distribution board. Record the value obtained. Repeat the test, measuring the neutral conductors and finally the current protective conductor; again noting the values obtained. Figure 14.9 clarifies this method.

2. Once the final ring circuit has been reconnected, select a suitable socket lying approximately at a mid-point position and short circuit both phase and neutral conductors to the protective conductor as illustrated in Figure 6.20.

3. Next, measure the resistance between phase and neutral conductors at the distribution board. For this test the fuse or circuit breaker must be removed from the circuit. The value obtained should be approximately

half of the original reading taken before the ring was complete.

4. To complete the test, measure the total resistance offered between both phase and protective conductor. This value should be close to the *sum* of one-quarter of the intitial reading taken across the phase conductors *and* one-quarter of the first value obtained from the open ends of the protective conductor as Figure 14.10 illustrates.

5. Disconnect the short circuiting arrangements and reinstate the socket outlet and protective device.

Insulation resistance test

Once all electrical work has been completed, an insulation resistance test must be implemented before any permanent connection is made to the supply. This is to verify that the insulation serving both cables and accessories is satisfactory and free from fault conditions existing between any current-carrying conductor and earth.

Larger installations may be divided into smaller groups for the purpose of testing, provided each fraction comprises not less than 50 points. The insulation value recorded should not be less than 1 MΩ.

As a practical example: an installation is served by four small distribution boards, each tested and found to be approximately 1.0 MΩ, each would collectively produce 0.25 MΩ. Because of this, large installations should be divided into smaller groups of not less than 50 outlets for the purpose of testing (see Chapter 2 for installation problems involving Ohm's Law).

The term 'point' is defined as any independent switch, socket outlet, fused connection unit and lighting point or a termination where fixed wiring yields to a permanently connected current carrying load. A twin socket outlet is defined as two points whereas a domestic cooker and control switch is described as one. This is illustrated more clearly in Figure 14.11.

Insulation resistance readings are measured using a DC source of not less than twice the nominal working voltage of the installation. The test voltage need not exceed 500 V for installations where the working voltage is not greater than 500 V or 1000 V DC where the voltage is above 500 V but under 1000 V.

Practical testing Testing may be carried out simply and safely by use of the many types of commercial instruments available today. Insulation testers fall into one of three categories:

1. *Handcranked insulation test meter.* A manually operated, hand-driven generator displaying up to three resistance ranges, developing a choice of 500 or 1000 V DC.
2. *Electronic analogue insulation tester.* This adopts the form of a push button, battery-operated meter, usually supporting up to three resistance ranges and a choice of three voltages. The test voltage output may be manually selected for use on either 250, 500 or 1000 V DC. The accuracy for this type of meter lies between ± 4−6 per cent of the full scale reading.
3. *Digital insulation tester.* This is a battery-operated electronic push button instrument, often incorporating just two resistance ranges. The lower scale enables readings to be made between 0 and 199.9 Ω and is ideal when installation fault-finding. The higher scale will measure resistance values from 0 to 199.9 MΩ. Readings on this scale are displayed in megohms or decimal divisions of megohms. The DC test output voltage may be selected for use on either 500 or 1000 V.

The accuracy factor for this category of tester is good, lying between ± 0.5−1.5 per cent of the reading obtained. Coupled with a large clear liquid crystal display panel, the digital insulation tester amounts to a sensible addition to an electrician's tool kit.

Figure 14.12 illustrates a proprietary digital insulation/continuity tester.

Testing procedure Before testing is carried out, remove or disconnect all lamps and current using equipment from the installation. Regulation 713-04-02 recommends that all electronic devices should be isolated from the supply in order to

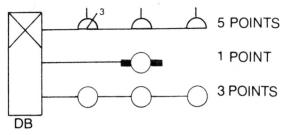

5 POINTS

1 POINT

3 POINTS

DB

Figure 14.11. The term 'point' may be defined as where fixed wiring yields to an electrical accessory connected to a current-carrying load.

Figure 14.12. Digital insulation/continuity tester. (*Reproduced by kind permission of Robin Electronics Ltd.*)

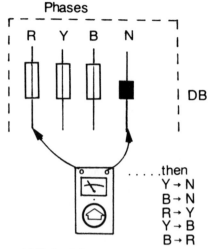

Figure 14.13. Insulation test between current-carrying conductors.

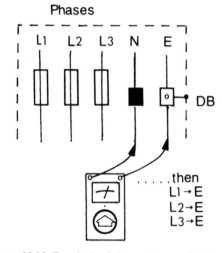

Figure 14.14. Test the insulation resistance offered between each independent conductor and the principal protective conductor.

avoid permanent damage from the test voltage. With all switches closed and fuses and circuit breakers made, the total resistance recorded must not fall below 0.5 MΩ when the nominal voltage is up to and including 500 V. Above 500 V up to 1000 V the minimum insulation resistance should not fall below 1 MΩ (Regulation 713-04-04).

1. First measure the insulation resistance between all current-carrying conductors. This should not be less than the values prescribed. Figure 14.13 depicts how a typical three phase and neutral installation is tested between each of the current-carrying conductors.

2. Next test the insulation resistance offered between each current-carrying conductor and the main protective conductor as shown in Figure 14.14. This again should not fall less than the values specified.

3. As a final check, verify the *collective* resistance offered between all current-carrying conductors made common as

outlined in Figure 14.15. Once again, the value obtained should not be less than the value defined.

4. Tested connections must be identified and labelled to meet specified requirements.

Extra low voltage circuits A minimum insulation resistance of 0.25 MΩ is required when testing SELV circuits, supplied from an isolating transformer. The circuit may be tested using 250 V DC test voltage. However, the *Wiring Regulations*, Table 71A requires a test voltage of

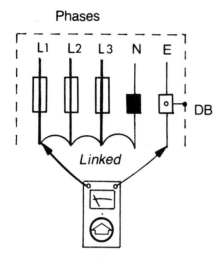

Figure 14.15. Insulation resistance is measured between grouped current-carrying conductors and the main protective conductor.

500 V to be placed between the SELV circuits and the low voltage supply circuit. The minimum insulation resistance recorded should not be below 5 MΩ.

When falling below the prescribed value Should readings obtained fall below the recommended value, a check must be made to establish the precise nature of the fault condition. Remedial measures should then be carried through in order to increase the level of insulation resistance to an acceptable level. Often the cause of low readings is the direct result of pinched or scuffed cables or damp contamination taking the form of plaster or building rubble.

Other reasons can stem from:

1. Faulty transformer (SELV) circuits.
2. Squashed cables.
3. Vandalism.
4. Wrongly connected conductors.
5. Poorly or incorrectly connected conductor.
6. Carelessly abandoned 'first fix' cables buried in the plaster line.
7. Nails inadvertently driven through a cable run.
8. Malicious damage.

Insulation of site-built assemblies
This regulation has been formulated to provide testing facilities for site-built assemblies and custom-built electrical equipment. As this test is of a special nature it need not be applied during the period of inspection reserved for other tests.

Briefly, this regulation requires all insulation constructed around or applied to live parts forming a site-built assembly to be tested. This is to verify that adequate protection has been afforded against direct contact to live conductive parts without the risk of flashover or breakdown occurring resulting from faulty or inadequate insulation or by the infiltration of solid foreign bodies or water. The test voltage must be equal to the voltage applied to similar factory-built equipment as when tested to *British Standards* 5486: 1977, a copy of which may be borrowed from any public library (Regulation 713-05-01).

A test voltage of 2500 V AC is applied to switch and control gear having a formal rated insulated value of between 300 and 600 V.

Protection by means of electrical separation
This method can be used as a means of protection against indirect shock with exposed conductive parts in accordance with Regulations 413-06-01 and 471-12. Testing should be carried out in the manner defined. Check that:

1. The secondary voltage is no greater than 500 V.
2. There is no physical continuity between the primary and secondary windings of the supply transformer.
3. The separated circuit is insulated from all other circuits and has a resistance value of no less than the value obtained between the primary and secondary winding of the supply transformer.
4. The installation is completely independent and is wired using a non-metallic sheathed multicore cable. Alternatively, insulated conductors may be drawn through PVCu conduit.
5. No live parts of the separated circuit are connected to any unseparated circuit.
6. All flexible cables are physically visible throughout their lengths.

SELC circuits SELC circuits supplied by means of an isolating transformer, motor generator or

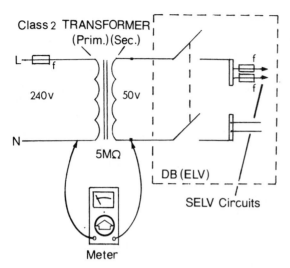

Figure 14.16. Continuity test verifying protection by means of electrical separation.

battery set must be thoroughly inspected in order to verify that electrical separation has been carried out. Alternatively the installation may be tested by means of a continuity meter, as schematically detailed in Figure 14.16. A minimum insulation resistance of 5 MΩ is required.

It is important that no interconnection exists between the circuit under inspection and any other electrical system. Complete separation is essential under the requirements of this regulation.

SELV circuits must be protected by means of barriers or enclosures when the nominal voltage exceeds 25 V AC or 60 V DC. Where this is impractical, low voltage circuits must be wired using conductors insulated to the maximum voltage present within the installation. Barriers and enclosures employed as means of protection in SELV circuits should be capable of withstanding a DC test voltage of 500 V AC for one minute.

Installations below 25 V AC or 60 V DC need not be protected against direct shock (Regulation 411-02-09).

Low voltage cable identification A simple but practical method of identifying grouped or bunched extra low voltage conductors from others of a higher voltage is by use of non-standard colours. Violet or white insulated cables would be ideally suited.

Protection by barriers or enclosures provided during the erection of an installation
Briefly, this regulation has been drafted in order to verify the presence of barriers or enclosures installed during the construction period of an installation.

Live or potentially conductive parts must be accommodated within suitable enclosures. The *Wiring Regulations*, Part 2, Definitions, defines the term 'enclosure' as: 'A part providing an appropriate degree of protection against certain external influences and a defined degree of protection against contact with live parts from any direction.'

Electrical accessories used as enclosures can include trunking and channelling, ducting and adaptable boxes. These are required to be strong and firmly secured when in place. Opening must be only by means of a tool or a key. Barriers used as flash guards, means of separation or fire security shields must be correctly fitted in an undamaged condition.

Preventing the spread of fire In order to prevent the spread of fire from one area to another, holes cut in walls and ceilings to allow for the passage of wiring must be sealed and made good using a suitable fire-resistant material once the installation has been completed (Regulation 527-02-01).

Electrical installations passing through walls and ceilings forming ducting, channelling or bus-bar trunking must be provided with internal fire-resistant barriers. These must be installed throughout the system where the installation passes from one area to another. Figure 14.17 illustrates how an armoured PVC insulated cable, clipped to the side of a concrete floor ducting, passing from one area to another is suitably provided with a fire barrier (Regulation 527-02).

Insulation of non-conducting floors and walls
In view of the special nature of this test and the strict supervision required, it is recommended that only a suitably qualified electrical engineer carry out the prerequisites of this regulation.

The test is designed to meet the requirements of a permanent but specialised installation where floor, ceiling and walls are electrically insulated

Substituting for known figures:

$Z_s = 1.2 + (0.0148 \times 20)$
$Z_s = 1.2 + 0.296$
$Z_s = 1.496\ \Omega$

Consulting *Table 41B1* of the *Wiring Regulations*: the maximum earth loop impedance permitted within a circuit served by an overcurrent protective device rated at 15 A to BS 3036 is 2.67 Ω.

Had the circuit been in the form of a radial, protected by a 20 A rewirable fuse, the value of Z_s would have to be reduced to 1.85 Ω, but in either case the total impedance calculated would meet the demands of this regulation.

Earth fault current Once the value of Z_s has been found, the earth fault current. I_f can be calculated by use of the following expression:

$$I_a = \frac{U_0}{Z_s} \qquad [14.2]$$

where U_0 is the nominal working voltage to earth.

Substituting figures,

$$I_a = \frac{230}{1.496}$$

$$I_a = 153.743\ A$$

In practical terms this would imply that a fault condition of negligible impedance occurring between phase and neutral, or from phase to earth, would meet the requirements of this regulation and be capable of rupturing the BS 3036 rewirable fuse within 0.4 seconds. Regulation 413-02-09 refers.

Consider the same practical example substituting the value of Z_s with 400 Ω! The maximum potential fault current would then be:

$$I_a = \frac{230}{400}$$

$$I_a = 0.5\ A!$$

In cases such as this, a residual current device would be installed to protect the installation.

Steel wire armoured cable to BS 6346 Table 14.2 provides approximate *Earth-Loop-Impedance* values for a selection of variously sized copper-phase

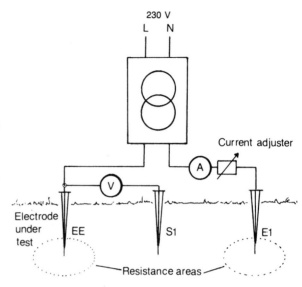

Figure 14.23. Earth electrode resistance — alternating current method.

conductors placed in series formation with the steel wire armouring protecting them; $(R1 + R2$, as in Expression 14.1). An allowance has been made incorporated within the table, for increase in temperature accompanying a fault current condition of negligible impedance to earth. Protective devices serving the circuit must never be higher than the current rating of the cable as damage to the cable and insulation could result.

Earth electrode resistance (if required)
There are two generally accepted methods of determining the resistance of an earth electrode in order to verify that its position taken does not alter the earth fault loop impedance to an unacceptable level.

Method A (an early traditional method) This employs a double wound low voltage transformer or hand-cranked generator in tandem with a high resistance volt meter. The final arrangement is put in series with a suitable scaled ammeter as illustrated in Figure 14.23.

The electrode under review, EE, must be completely disconnected and physically separated from the installation served. An auxiliary electrode, E1, is then placed at a distance which

TABLE 14.2   Value of $R1$ and $R2$ for Steel Wire Armoured cable

Cross-sectional area of phase conductor. (mm^2)	Value of $R1$ and $R2$ (Ohms per kilometer)		
	Two-core SWA cable	Three-core SWA cable	Four-core SWA cable (All conductors of an equal size)
1.5	30.3	28.3	26.8
2.5	22.0	20.6	19.0
4.0	16.1	14.74	11.42
6.0	13.4	9.87	8.90
10.0	7.62	7.02	6.49
16.0	6.05	5.40	4.16
25.0	4.42	3.97	3.48
35.0	3.89	3.33	2.97
50.0	3.31	2.90	2.08
70.0	2.88	1.99	1.81
95.0	1.99	1.77	1.44
120.0	1.80	1.59	1.06

ensures that its earth resistance area will not overlap or influence the resistance area of the electrode under test. An ammeter is wired in series formation between the auxiliary electrode and the power source. A supplementary electrode, S1, is staked at a mid-point position lying between EE and E1, and the voltage between EE and S1 noted.

It is important that the two resistance areas do not overlap. This may be confirmed by moving the supplementary electrode 'S1' an equal distance of 5 or 6 m from the mid-point position and retesting.

If after calculation the three values obtained are similar to each other, it may be said that the resistance areas are separated from each other. The average resistance of the electrode under test can then be calculated by use of the following expression:

electrode resistance =

$$\frac{\text{average voltage between EE and S1}}{\text{average current between EE and E1}} \quad [14.3]$$

When unacceptable differences exist between the three independent tests, the auxiliary electrode must be staked at a greater distance from the electrode under review and each test repeated.

Stray currents within the soil due to supply distribution leakages can cause difficulties when employing this method of testing.

Method B (testing by proprietary means) This will take the form of either a hand-held driven generator or a battery operated electronic device supplying a working frequency of up to 800 Hz.

Testing is carried out similarly to the traditional method. The electrode and auxiliary electrode are staked at least ten times the maximum dimension of the electrode system and connected to the test equipment. A supplementary electrode is positioned mid-way between the two staked electrodes and is also connected to the meter.

This is an ideal method of measurement as the resistance value of the electrode under test may be taken directly from the scale of the test equipment.

As with Method A, to verify the intitial reading, the central supplementary electrode must be removed and reinstated first further and then nearer to the electrode under test, keeping the distance travelled from the central point constant in both directions. Each supplementary electrode move *must* be accompanied by an earth resistance test. Once three similar consecutive readings have been gained, an average figure can be calculated to represent the value of the electrode under test.

Figure 14.24 schematically outlines the method of testing using a commercial earth resistance meter. When hand-generated equipment is used, any stray residual currents present within the soil will cause the needle to oscillate. This can be

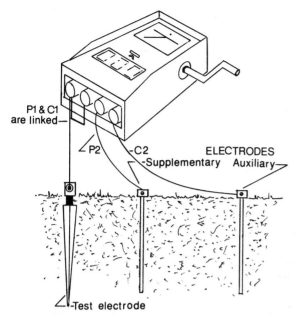

Figure 14.24. Measuring earth electrode resistance by means of a proprietary test meter.

Figure 14.25. Digital residual current device tester. (*Reproduced by kind permission of Robin Electronics Ltd.*)

offset by either cranking the generator faster or slower than the speed at which the meter is being driven. This method of testing may be found in BS 7430: 1991.

Testing the operation of a residual current device
This test has been designed to verify the automatic and mechanical effectiveness of a residual current device and is implemented by placing a nominal load between the phase and protective conductor on the consumer's side of the installation.

The test must last no longer than one second, but ideally the tripping current should cause the device to activate within 0.2 seconds, disconnecting the load from the supply as it opens. (Regulation 713-12-01).

When a residual current device is employed as additional protection against direct contact, its residual operating current should not exceed 30 mA and provide a tripping time of 0.4 seconds at a residual current of 150 mA. Should this test prove unsuccessful, then the duration of the test current is limited to 0.5 seconds. Regulation 412-06-02 and BS 4293: 1983 refer.

Residual current device testers There are many professional test meters to choose from to suit all pockets. These are known as *digital RCD testers* (Figure 14.25), and provide for residual tripping currents from 5, 10, 30, 100, 300, 500 mA. Often there are facilities to verify the tripping action of a 30 mA device when required to activate disconnection within 40 ms at a residual tripping current of 150 mA.

Practical testing To meet the requirement of this regulation a suitable load, such as a safety lead light, should be connected to one side of a twin socket outlet. A test meter is then plugged into the remaining vacant outlet.

The effectiveness of the RCD is assessed by pressing the test button and recording the time in milliseconds for the device to activate. For greater accuracy an integral change-over switch provides a choice of 180 degrees so that testing may be started on the positive or negative half cycle. The majority of commercial meters meet with an accuracy of between ±3−7 ms.

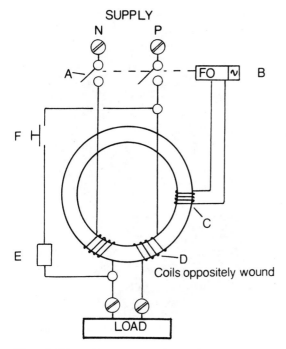

Figure 14.26. Residual current device (RCD): A, double pole switch; B, tripping relay; C, fault-sensing coil; D, toroidal transformer and induction coils; E, test resistor; F, test button.

systems is prescribed in Chapter 73 and Regulations 731–744 of the *Wiring Regulations*. It is advised that inspection be carried out at intervals between 3 and 60 months depending on the nature of the installation.

Regular periodic inspections for specific types of installations are recommended by the Institution of Electrical Engineers as follows:

1. General electrical installations — 5 years.
2. Agricultural installations — 3 years.
3. Churches (new installations) — 2 years.
4. Caravan site installations — 1 year
5. Cinemas and theatres — 1 year (mandatory requirement).
6. Fire alarm systems — 1 year (mandatory requirement).
7. Launderettes — 1 year (mandatory requirement).
8. Petrol filling stations — 1 year (mandatory requirement).
9. Temporary electrical installations serving building sites — 3 months.
10. Swimming pools — 1 year.

Once the automatic disconnection time has been satisfied, the effectiveness of the manual test button, incorporated within the current device, must also be verified. At times a faulty current device will trip when a fault current flows but will fail to respond to the manual test button facility. Alternatively, should an instrumentation test prove inconclusive or puzzling, the tripping action may be tested by pressing the manual test button. If successful, the problem could stem from a faulty phase or protective conductor.

Figure 14.26 outlines the internal wiring arrangement in schematic form of a typical residual current device connected to a simple external load. As a secondary check, a digital clamp meter can be placed in circuit to record current flow through the outgoing phase conductor. This will indicate that disconnection has taken place should an incandescent lamp be unavailable.

Periodic inspection

The periodic inspection and testing of electrical

After testing

Once re-inspection and testing have been satisfied, a certificate must be issued, verifying that the installation is electrically safe and complies with the general requirements of the *Wiring Regulations*. This should be signed by a suitably qualified electrical engineer or a person of technical competence, but any defects or omissions should first be made good.

Copies of this form may be obtained from the Institution of Electrical Engineers or reproduced privately in the style of the sample provided within the *Wiring Regulations*. If this method of reproduction is adopted, please remember to provide the usual courtesy acknowledgement.

On completion of work, a notice should be placed near or attached to the main distribution centre recommending that the installation be periodically tested and inspected as prescribed within the *Electrical Wiring Regulations*. This should be accompanied by the date of the last inspection, together with recommendations for a

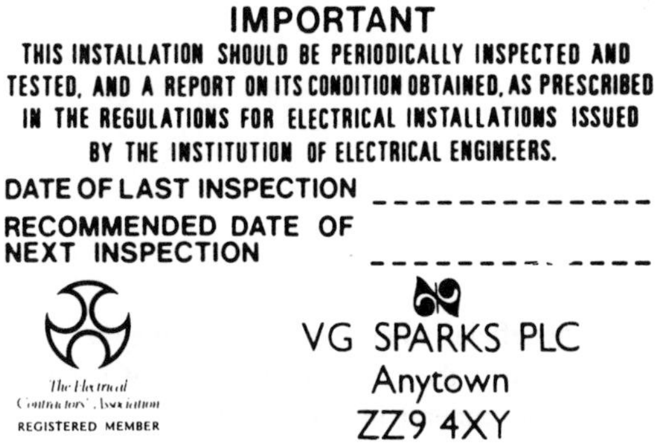

IMPORTANT

THIS INSTALLATION SHOULD BE PERIODICALLY INSPECTED AND
TESTED, AND A REPORT ON ITS CONDITION OBTAINED, AS PRESCRIBED
IN THE REGULATIONS FOR ELECTRICAL INSTALLATIONS ISSUED
BY THE INSTITUTION OF ELECTRICAL ENGINEERS.

DATE OF LAST INSPECTION _ _ _ _ _ _ _ _ _ _ _ _

RECOMMENDED DATE OF
NEXT INSPECTION _ _ _ _ _ _ _ _ . _ _ _ _

VG SPARKS PLC
Anytown
ZZ9 4XY

The Electrical
Contractors' Association
REGISTERED MEMBER

Figure 14.27. Periodic inspection and testing notice.
(Reproduced by kind permission of Systems and Electrical
Supplies Ltd.)

future inspection. Figure 14.27 illustrates this
notice in full together with the minimum sized
characters recommended.

Summary

Appropriate testing must be carried out relevant to
the installation. There are tests which are not
applicable. It would be hardly suitable to conduct a
test verifying the insulation value of non-
conducting floors and walls in a domestic
installation!

Because of this, testing must always be both
selective and thorough. It might be tempting to
carry out a quick insulation test before
commissioning but this approach is both slovenly
and potentially dangerous. Formal verification and
a detailed examination are essential once an
installation has been completed.

Always maintain a good standard of
workmanship and keep within the
recommendations and guidelines as laid down in
the *Wiring Regulations* or *National Electrical
Codes of Practice*.

Remember — good electricians are a valuable
asset!

Handy hints

* A check should be made on all test
  equipment before an electrical inspection is
  carried out. Recalibrate if necessary.
* Use rubber gloves when earth electrode
  resistance tests are carried out, as a
  precaution against shock.
* Electrodes designed for use with a fault
  voltage-operated circuit breaker must have a
  working resistance of below 500 Ω.
* It is unwise to fault-find by use of a
  screwdriver which has incorporated a neon
  voltage indicator lamp. Such screwdrivers
  are capable of registering stray static and
  inductive currents and therefore are not
  reliable. Always use the correct test
  equipment to suit the task in hand.

Part 2 review questions

Listed are 16 questions to test your new-found practical skills in electrical installation work. Answers are given in Appendix A.

1. In the UK, supply authorities have to comply with the *1937 Electrical Supply Regulations*. Name four earthing arrangements associated with power distribution.
2. Main equipotential bonding conductors must connect all extraneous and exposed conductive parts. Explain the term *extraneous conductive parts*.
3. There are several different methods used when wiring a lighting circuit. List four ways how this might be carried out.
4. Why do modern kitchens justify an independent final ring circuit?
5. There are several good reasons why solid drawn steel conduit is chosen to form an installation. Itemise three of them.
6. Extremely low temperatures can cause PVCu conduit to rupture when being formed into a set or right angle. How might this problem be overcome?
7. Because of its unique construction and versatility, mineral insulated cable is ideally suited for wiring tasks demanding a high degree of reliability. Suggest three such types of installation which would prove beneficial when wired in this way.
8. What is the main disadvantage of using old style mineral insulated cable?
9. Name two installation methods of laying armoured PVC insulated cable.
10. Removing overtight immersion heater elements can be a problem when they fail to respond to conventional methods. Describe briefly how such a problem might be overcome.
11. There are two generally accepted methods of wiring a domestic central heating control system. Name one of them.
12. List two types of capacitors used in low frequency electrical work.
13. An alternating magnetic field produces eddy currents. These residual currents are generally unwanted as they generate wasteful energy in the form of heat. Describe how this problem is virtually overcome.
14. Define the running characteristics of a capacitor start induction motor.
15. After two years of continuous running, motor bearings should be removed and washed thoroughly. Why is it unwise to pack the bearings tightly with grease?
16. How many instrumental tests are recommended to be carried out once an installation has been completed?

A summary of theoretical expressions used in Part 2

[11.1] cross sectional area of a conductor $= \pi r^2$

[14.1] $Z_s = Z_e + (R1 + R2) \times C$

[14.2] $I_a = \dfrac{U_0}{Z_s}$

[14.3] electrode resistance $=$

$$\frac{\text{mean voltage between EE and S1}}{\text{mean current between EE and E1}}$$

For a list of abbreviations please see p.xi at the front of the book.

Part 3   Health and safety at work

Accidents, no matter how small, are a constant and nagging reminder that we are not being as careful as we should. In 1983 alone, there were 1612 registered electrical accidents where hospital treatment was required and between 1977 and 1982 there were 1033 deaths caused by faulty electrical appliances. As well as being a disruptive nuisance to everyone involved, accidents are a significant contributory cause for lost man-hours in our industry today.

The third and final section of this book takes a fresh look at the major reasons of worktime misadventure, and is written to help the reader develop an awareness in order to guard against the errors of judgement which often accompany over-familiarity or inexperience. For those who are less skilled, Part 3 will be equally beneficial and will show how to learn from the mistakes of others.

This section is not intended to be a complete comprehensive guide to the safe use of electricity but a practical insight and introduction to sensible working practices within the electrical installation industry.

Always be aware that electricity can kill.

Accidents can be prevented by practising vigilance, developing an awareness to potential hazards, keeping to the rules and not applying your own. Safety is an attitude of mind which we should all try to cultivate. Accidents do not just happen; they are caused by lack of care and understanding.

Remember — the responsibilities carried at work will not be only to yourself but will also extend to your companions and associates. If the advice and guidelines appearing in this section are heeded, then hopefully serious accidents stemming from carelessness or lack of knowledge may be averted or minimised.

Part 3 has been fashioned to be used as a 'dip-in' section in order to provide instant guidance or advice appropriate to the task in hand and meets with the requirements of *National Vocational Standards* for the electrical installation engineering industry and the revised *City and Guilds GCLI 236 syllabus*.

15 Safe working practices and general guidelines

In this chapter: Common causes of accidents. General dos and don'ts. Installation guidelines. Authoritative bodies. Agricultural installations. Working with primary and secondary cells. Disconnection procedures. Installing and dismantling access platforms.

Common causes of accidents

Study has shown that the most common causes of accidents at work are:

1. Carelessness.
2. Lack of knowledge.
3. Human limitation.
4. Fatigue and listlessness.
5. Horse-play.

Carelessness at work is one of the major causes of accidents in industry today. It can stem from a seemingly innocent act of day-dreaming or being preoccupied with personal concerns through to more serious problems such as tiredness, the after-effects of alcohol and drugs or from an adoption of a cavalier attitude to the work in hand. Whatever the cause, should uneasiness be felt or a personal risk be involved, get assistance. It will be a far safer decision.

Accidents occurring through lack of technical knowledge are generally caused by operatives who are told to carry out tasks for which they are unqualified. Often a great deal of pressure is placed on the young by an employer in the name of cost-effectiveness to complete a job, irrespective of how experienced or qualified the operative might be.

Human limitation frequently defies good common sense. We are inclined to believe we are capable of far greater physical achievements than we really are and it is only later that we become aware that all is not quite as it should be. Strains and muscular disorders are responsible for many lost man-hours in the electrical industry — time that need not have been lost if common sense had prevailed.

Long hours, working weekends and insufficient sleep are the major factors responsible for fatigue and listlessness. The insatiable desire for extended productivity can often be a major contributory cause for carelessness and preoccupation throughout the working day. Inevitably this leads to mistakes being made; mistakes which can lead to unfortunate accidents. Try to achieve a sensible balance between work and recreation in order to reduce the stress and tiredness factor which frequently accompanies extended overtime.

A working environment can often be a tempting venue to carry out high jinks and practical jokes on unsuspecting workmates and associates. This will happen when discipline is relaxed and general site and working conditions unpleasant. Wise leadership, together with a clean and tidy site, will provide a sound foundation for a more favourable and safer workplace.

A tidy site is a safe site

It should be in everyone's interest to keep all work areas free from clutter and as unobstructed as possible. Remove all offcuts from the floor and place them in a suitable container for collection at a later date. Materials that could possibly cause or add to a fire hazard should always be stored in a locked room with two key holders. On large building sites it is easy to inadvertently block passageways and exits with stored material and equipment which is required for immediate use. Leave a space free from tools and accessories at least half the width of a passageway so evacuation may take place in the event of an emergency occurring.

Protective clothing There seems to be a growing tendency amongst electrical workers to abandon the wearing of overalls during the milder seasons. Overalls not only shield everyday working clothes from unnecessary wear and contamination but also provide a reasonable degree of protection from the many hazardous working environments which may arise from time to time. Always wear the correct protective clothing to suit the task in hand. It makes good common sense.

General dos and don'ts: your responsibilities

Do

1. Obtain professional advice when dealing with asbestos.
2. Report faulty drills, leads and general plant to your site supervisor.

3. Maintain personal tools to a high standard. Misused and neglected tools are potentially dangerous. Wipe relevant tools with a lightly oiled rag to protect and clean. Rusting not only can cause damage but may also be responsible for many sorts of medical problems if introduced into an open wound.
4. Report defects in safety equipment to your supervisor.
5. Wear good strong boots or protective shoes on building sites (Figure 15.1). Unprotected trainers might be cheaper, and possibly more comfortable, but will not give the protection required when working in such conditions.
6. Be aware of other trades working overhead. Wear suitable protective clothing, including a hard-hat.
7. Check that the ladder you are to use is in no way damaged.
8. Ensure that your ladder is positioned on a

Figure 15.1. It is wise to wear high-performance safety footwear conforming to BS 5750 whenever work is carried out on a building site.

consequence of mice and rats nibbling cable. Rubber and PVC insulated conductors should be clipped in a way which discourages rodents from gnawing them. Regulation 522-10-01 refers.

16. Capacitance can often be stored in long cable runs whenever insulation testing is carried out. To avoid an unnecessary and unexpected shock, short circuit the conductors and current protective conductor after each test. Physical reaction to a capacitance shock is often far more severe than the shock experienced. This could be hazardous if, for example, testing is carried out whilst standing on a pair of steps or on the rungs of a high ladder.

17. If fluorescent fittings have to be tested, discharge any built up capacitance by short circuiting the phase and neutral conductors after each insulation test (Figure 15.7). Some fittings are manufactured with a power factor capacitor in circuit.

18. Never play with stored capacitors.

19. Masonry cable clips have been known to snap and fly, causing physical harm, if not squarely hammered home. Damage to the eyes can be avoided by wearing suitable eye protection.

20. Some luminaires, because of their design features, are unsuitable for mounting on flammable surfaces such as chipboard, hardboard or timber. Always check the manufacturer's recommendations before installing.

21. Wear an appropriate face mask when working in roof spaces which have been lined with a layer of fibreglass insulation. The older the insulation, the greater the risk of loose or airborne fibres.

22. Dress in suitable clothing, such as overalls, steel capped boots or shoes, ear and eye protectors and a hard-hat. Kit yourself out relative to the task in hand. You might feel a little foolish, but it could help to save your life.

23. Large cable drums must always be manned when dispensing cable. To create both easier and safer working conditions, cable should be drawn from the top of the drum as illustrated in Figure 15.8.

24. Care must be taken when handling wooden cable drums so as to avoid splinters, nails and sharp obstacles. It is advisable to wear industrial gloves.

25. Cable drums stored outside for any period of time should be physically inspected to ensure that no structural deterioration has taken place. Manhandling a large disintegrating cable drum could prove to be quite an alarming experience, so check first; it could prevent injury.

26. Grease applied to a cable drum spindle will enable the cable to be dispensed with a greater degree of control. The use of well-maintained cable rollers will reduce the amount of physical effort required when manhandling cable from the drum and will help keep the cable free from damage.

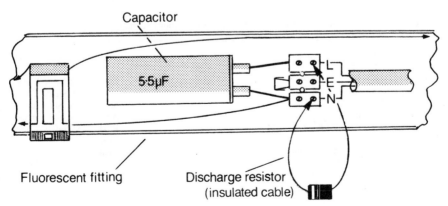

Capacitor

5·5μF

Fluorescent fitting   Discharge resistor
                             (insulated cable)

Figure 15.7. Discharge capacitance after each insulation test.

Figure 15.8. It is both easier and safer to dispense cable from the top of the drum.

27. Cables used in an installation must be as specified and suitable for the intended use. It would not, for example, be safe to use a flexible rubber cable underground where an armoured PVC insulated cable had been requested.
28. Cables must suit environmental conditions. Regulation 522 confirms.

Authoritative bodies

Installation practice will vary depending on environmental conditions and the nature of the work to be carried out. Mandatory (compulsory) and statutory requirements covering many classes of installation are published and easily obtainable. Listed in Table 15.1 are a selection of recognised statutory and advisory publications covering a wide range of safe installation practices.

Handy hints

- Avoid using power tools in the rain.
- To prevent overheating, extension leads should be fully unwound from the reel.
- Fit a plug-in type residual current device to protect against electric shock when using an extension lead.
- Handles may be made for files by filling a suitable length of metal conduit with timber. Once firmly embedded, the timber insert can be drilled to accommodate the file shaft as illustrated in Figure 15.9.

Agricultural installations: avoiding accidents

Agricultural installations can, by their very nature, produce many hazardous environmental conditions where extra care is necessary. Many accidents are preventable simply by using greater vigilance. A

TABLE 15.1 A selection of statutory and advisory regulations serving the electrical industry

Item	Title of publication	Authoritative body
1*	IEE *Wiring Regulations*	The Institution of Electrical Engineers
2	Quarries (Electical) Regulations, 1956	Health and Safety Commission
3	Electricity Supply Regulations, 1937	Secretary of State for Energy
4	Agricultural Regulations, 1959	Agricultural and horticultural installations
5	Factories Act, Special Regulations, 1908 and 1944	Health and Safety Commission
6	Cinematograph Regulations	Home Office
7	Coal and other mines (Electricity) Regulations, 1956	Health and Safety Commission
8	Electricity at Work Regulations, 1989	Secretary of State for Energy
9	Memorandum of Guidance on Electricity at Work Regulations, 1989	Secretary of State for Energy
10	The Deposit of Poisonous Waste Act, 1972	UK Parliament

* BS 7671: (1992); advisory

Figure 15.9. An emergency file handle can easily be made from a small piece of steel conduit and timber insert.

selection is given of the more important safety aspects which should be observed.

1. Storm-soaked wooden distribution poles often carry a dangerous electrical potential to earth. Standing on a metal ladder which is leaning against a rain-drenched pole can be extremely unsafe and should be avoided, no matter how urgent the job. Always isolate the current before any work is carried out.

2. Current leakages to earth from storm-soaked overhead conductors, via wet and contaminated line taps and insulators, through to rain-saturated wooden distribution poles, are often responsible for activating the principal residual current device serving an installation. This may be seen more clearly by studying Figure 15.10. The more distribution poles there are, the greater the current drain to earth. Animals standing in a field close to the poles could respond to leakage current as Figure 15.11 shows.

3. Care must be taken when carrying long metal ladders in agricultural premises to avoid touching overhead cables. Older installations are often served with a network of bare hardened copper conductors supported by insulators on wooden distribution poles (Figure 15.12). Regulation 522-06-01 refers.

4. Test voltage- and current-operated circuit breakers regularly, especially after an electrical storm.

5. Before commencing work, inspect all power tools and extension leads for damage. For additional safety, plug in a portable residual current device for protection against electric shock. Never add extra cable to an extension lead using connectors or taped joints if the original lead does not meet with your requirements. In hazardous conditions, such a makeshift joint is potentially dangerous.

6. Use only 110 V equipment when working on-site; it could save your life!

7. Thoroughly bond milking parlours and dairies. Check that the installation is served with a sensitive residual current device. Livestock have been known to suffer lethal or disabling shocks from as low as 25 V when conditions are wet and contaminated.

8. Always use the correct fuse wire rating; better still use a suitable circuit breaker (re Regulation Table 53A).

9. Protect all lamps and machines from any potential environmental danger. Ensure that all safety guards serving machines are securely fixed before the completed installation is handed to the client.

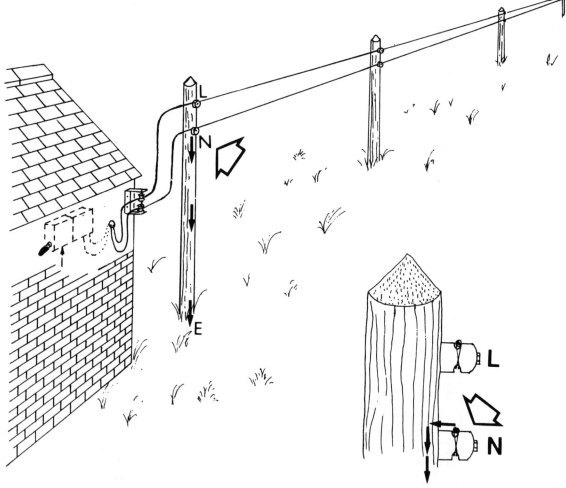

Figure 15.10. The first three poles illustrated have phase to earth insulation values of 2000, 2500 and 50 000 Ω respectively. The total current leakage to earth, when served with 230 V, would amount to 211.6 mA.

10. Large amounts of water are used to wash down and clean milking parlours and dairies; it is unwise to use block connectors in-line to serve current-consuming equipment. If an in-line connecting device has to be employed in a flexible cable, as with a milk paddle motor serving a bulk tank, it is more prudent to use a waterproof snap-on connection unit to provide a greater margin of safety.

11. Always wear overalls or well-fitting protective clothing to suit the working conditions. It would hardly be appropriate to wear a sweatshirt, trainers and jeans when employed in an agricultural installation.

12. In some environments it is necessary to wear a suitable face mask, for example, when work is being carried out in a grain storage silo, poultry farm or piggery.

13. By their very nature, agricultural installations can harbour many unpleasant, dangerous germs. As a precaution against unnecessary sickness, always wash your hands thoroughly before eating.

Handy hints
- Avoid day-dreaming; it could lead to misadventure.
- Never fool around with fully charged capacitors.
- Do not use open fronted quartz halogen

The stability of such a structure is greatly dependent on devices known as *scaffolding ties*, which are attached with clamps to tubular poles at distances not more than 8.5 metres in either a vertical or horizontal direction. When ties have to be removed, for example, to install a wall-mounted luminaire, it is important that supplementary ties are positioned and fixed to replace the tie that is to be removed. Again, never attempt this yourself but ask somebody who is qualified to carry out any necessary alterations.

Proprietary unit scaffold

This type of scaffolding, known as *prefabricated aluminium alloy tower* or *proprietary modular scaffold* is assembled from steel or aluminium sections and frames on site. Its purpose, like traditional scaffolding, is to support access platforms at various heights enabling work to be carried out safety.

Components Modular scaffolding is designed to be site assembled with use of the following slot-in or spring-clip components.

1. End frames (1.3 m or 0.75 m wide).
2. Adjustable legs with wheels or base plates.
3. Horizontal and diagonal braces (installed every fourth vertical metre), Figure 15.18.
4. Stabilisers and outriggers for additional strength and stability, Figure 15.19.
5. Safety guard-rails and toe-boards. (The uppermost guard-rail must be positioned between 1.0 m and 1.05 m from the working surface of the platform.)

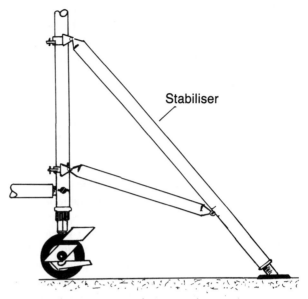

Figure 15.19. Stabilisers are fitted for additional strength and stability.

6. Access platforms, with or without a hinged hatch.
7. Internal access ladders.
8. Warning notices.

When choosing a prefabricated aluminium alloy tower it is advisable to purchase from a reputable manufacturer, such as *Access International*, or alternatively from a supplier of free-standing towers who is a member of the *Prefabricated Aluminium Scaffolding Manufacturer's Association* (PASMA). All PASMA member companies both design and build their products to meet the requirements of BS 1139, Part 3: 1994 and the European directive HD 1004.

Erecting an access platform

The following safety points and guide-lines should be considered whenever a free-standing access tower is to be assembled:

1. Ensure that all components are free from damage and are of the same design.
2. Check that the ground is firm enough to assemble an access tower and that work may be carried out safely without danger to yourself or to others. Read the instruction

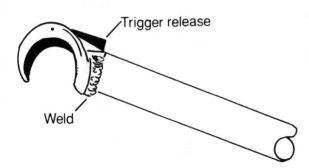

Figure 15.18. Automatic self-priming brace hook.

manual. Keep the work area tidy and free from obstacles. Cover all pit holes with a suitable material. Wear a hard-hat.

3. If the working area is virgin ground, provision must be made to prevent the adjustable legs from sinking into the soil. Building planking may be placed on the ground to prevent such a problem from happening. Castors should not be used. Use base plates.

4. If the working area is hardened soil which has been previously levelled, base plates should be attached to the adjustable legs of the tower. It is not wise to fit castors at this stage as problems can arise when pushing the tower across seemingly level ground. Wheels are ideal for paved areas but when used check that the castors are locked when the tower is being assembled.

5. Always remain faithful to the installation instructions, especially when installing diagonal braces, guard-rails, toe-boards and tower stabilisers. The latter are used in circumstances which require the tower to be assembled higher than recommended by the makers. Ensure that the principal base sections are level before building work continues, Figure 15.20.

6. Tie the tower to an adjacent structure. When this is impractical, secure by means of spikes or ground anchors. If additional stability is required, use ballast weights, available in 25 Kg units. These should be hooked to the lowest horizontal rung of the tower, Figure 15.21.

7. Keep tools in a safe place. A screwdriver dropped from an overall pocket from a height of perhaps 10 metres, could be the cause of personal injury to an operative working directly below.

8. Slot one section into another with care as illustrated in Figure 15.21. If force is required the component is probably damaged. It is important to use diagonal and horizontal braces wherever appropriate or recommended within the manufacturer's instruction manual.

9. Provide an intermediate rest platform at approximately every four vertical metres supporting a hinged hatch, Figure 15.22. A guard-rail and toe-board should also be fitted at this point.

10. Internal access ladders may be fitted so that the direction of climb is always opposite to the two adjacent ladders either side of the ladder being used. Alternatively a fixed internal vertical ladder may be used as illustrated in Figure 15.22.

11. If practical, fit a hand-rail to serve the access ladder for additional safety.

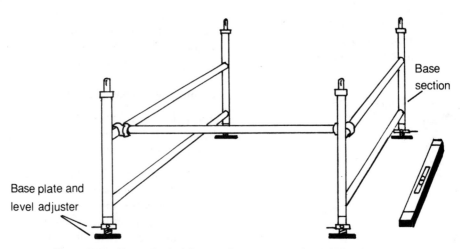

Base section

Base plate and level adjuster

Figure 15.20. Ensure that both base sections are level before building the tower.

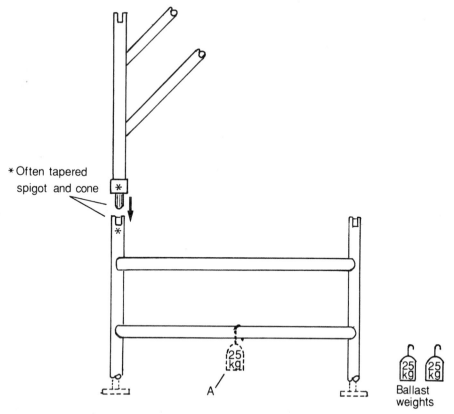

Figure 15.21. Slot one section into another with care. (A, Ballast weights may be used when additional stability is required once the tower has been completed.)

12. Working platforms are made from an aluminium or steel frame decked with non-slip plywood. The platform is supported firmly on the scaffold by means of hooks or *spring couplings* housed within *half-cups* of moulded aluminium welded to each corner of the metal frame. The moulded half-cups are snap-fitted to the horizontal scaffolding. When machine tools and material items are to be left within the working area, two platform widths must always be used.

13. Once the principal working platform is in place a safety guard-rail together with a set of toe-boards must be installed. The safety rail, which may be doubled, will help to prevent personal misadventure, while the toe-boards will stop tools and plant from being nudged off the access platform during the course of the electrical installation, Figure 15.22.

14. The wheels of a mobile tower must always be locked before any work is carried out. Operatives, materials and tools must not be on the tower while it is in motion. It is very tempting to hitch a ride as it obviously saves both time and energy but it only needs one mishap to cause an accident. Never move the tower by mechanical means; always push or pull it.

Dismantling an access platform

The manner in which dismantling is carried out is the reverse of the erecting procedure.

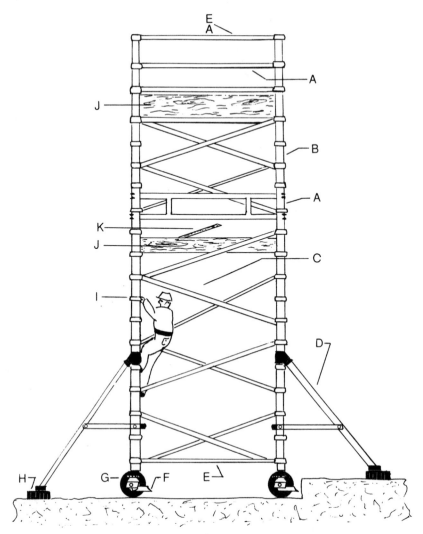

Figure 15.22. An access tower — side view. A, safety rail; B, end frames; C, diagonal braces; D, stabiliser; E, horizontal brace; F, castor lock; G, castor; H, friction pads; I, internal ladder span; J, toe-board and platform; K, trap-door and platform.

Never attempt to dismantle on your own; always work accompanied by a competent colleague.

The use of protective head covering and suitable safety footwear is essential.

The following points are useful practical guide lines to consider:

1. Start from the topmost handrail. Dismantle one section at a time. Components which are difficult to remove should be gently tapped with a small lump hammer at the problem end. Tapping upwards will release the component.

2. Use a strong nylon rope to lower each section and component to a ground-based colleague to untie and stack.

3. Throwing components to the ground is not only dangerous but could cause damage to the scaffolding sections and if the mobile is hired, the hire company could claim for compensation.

4. Never remove diagonal braces until such

time it is required to do so. This will prevent the tower structure from collapsing under its own weight. Keep the wheels locked.

5. Stack the separated components neatly on a wooden pallet for transportation and try to avoid excessive handling which can amount to wasted time. Clean and lubricate where necessary.

6. A splash of coloured paint on each component will help identify your company's modular scaffolding from others of a similar pattern.

7. Maintain the cleanliness of the working area.

Other means of access

Other types of industrial access platforms such as hydraulic, scissor (sometimes known as *cherry-pickers*) and mast-supported platforms are usually very expensive to hire. However, the high cost is often justified when high and awkward positions have to be accessed when compared to the cost of traditional scaffolding arrangements.

Safety points

The following points should be considered whenever work is undertaken using prefabricated access and working towers:

1. Read and understand the instruction manual.
2. Check for damaged components and make sure that the contents are of the same design and are in working order.
3. Assemble on a firm and level base. Use wooden sole boards when considered necessary.
4. Stabilisers must be positioned to provide a height to base ratio of 3 : 1 when the tower is used outside. Internally, a height to base ratio of 3.5 : 1 may be maintained.
5. Ensure that the ground is capable of supporting the tower.
6. Toe-boards and guard-rails should be fitted to all platform areas above 2 metres from the ground.
7. Towers over 10 metres in height must be secured to an adjacent fixed structure.

8. Castors must be mechanically locked before work is carried out from the tower.
9. Do not use an external ladder to gain access to the top working platform. Only climb the horizontal scaffolding rungs if they are evenly spaced. Climb internally if a ladder arrangement has not been included in the design of the tower.
10. Never move a tower by mechanical means; only by pushing or pulling at, or near to, the base area.
11. Keep aluminium components away from corrosive material.
12. Do not exceed the maximum recommended working load or height. Care should be taken when using percussion tools.
13. Keep the tower away from ground and overhead obstructions.
14. Never use a step-ladder from the uppermost platform.
15. Towers should not be used during periods of high wind. Tie the tower to an adjacent structure when used in exposed conditions. Do not place plastic sheeting around the tower; it will act as a wind sail.
16. A warning notice should be left on the tower if structurally incomplete.
17. Take extreme care in public areas. Tie the tower in to an adjacent structure if to be left unattended.

Further reading

British Standards 1139, Part 3: 1994.
Specifications for prefabricated access and working towers.
European Harmonisation Document, HD 1004.
Health and Safety Executive: Tower Scaffold Guidance Notes GS 42: 1987.
PASMA Operator's Code of Practice (4th Edition)

Handy hints

- Static voltages can be stored within the body and can become a source of annoyance during very dry weather. Nylon carpets and man-made rubber-soled footwear are a contributory cause to this phenomenon. Should it prove a nuisance when electrical work is being carried out,

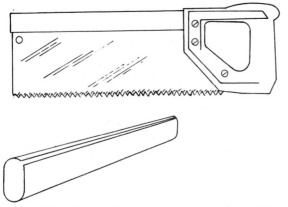

Figure 15.23. Oval conduit cut lengthways will make a reliable hand-saw blade protector.

discharge yourself to earth through a neon test screwdriver every now and then.

- Oval plastic conduit cut lengthways can make excellent shields to protect the blades of wood chisels and handsaws. Figure 15.23 illustrates.

- Avoid pushing a prefabricated installation tower across rough ground. It can be extremely difficult and potentially unstable. Always keep off the tower whilst in transit.

- Store steel conduit undercover or indoors during cold weather. Freezing temperatures can cause frost burn when the conduit is manhandled.

16 Personal hazards

In this chapter: Fire hazards, Fire extinguishers. Lifting techniques. First aid, dealing with electric shock. Mouth-to-mouth resuscitation.

Fire hazards

Accidents involving fire caused by power tools or equipment are often attributed to the people involved. Failure to respond to fault conditions through pressure of work or lack of experience contributes towards many misadventures. Remember that prevention is far better than cure — for lives and property are at stake when fire strikes.

Listed are a random selection of the more important safety aspects which should be observed.

1. Familiarise yourself with the appropriate action to be taken in case of fire.
2. Should work be undertaken in a large establishment, get to know where the fire call points, extinguishers and emergency exits are located.
3. On discovering a small fire:

    (a) sound the alarm;
    (b) try to extinguish the flames;
    (c) do not endanger yourself;
    (d) if the fire cannot be put out, escape by using the safest route out of the building.

4. Respect 'no smoking' areas. No smoking means NO SMOKING.
5. Never smoke in a thatched roofing space.
6. Current-consuming equipment, such as powered television amplification units, should never be installed in a loft which is served by a thatched roof.
7. Cable burning must only be carried out using a suitable incinerator. In the UK it is against the law to burn the insulation from cables under open site conditions for it will add to environmental atmospheric pollution.
8. Ensure that all fuses and circuit breakers are of the correct type and size and suitable for the requirements of the circuit.
9. Never overload a temporary installation designed to serve a building site. Conditions are often harsh, so check the wiring and all current-consuming equipment at least once a month in order to meet the requirements of the electrical regulations.
10. The use of an open-fronted quartz halogen hand lamp is not recommended as an aid when carrying out agricultural or horticultural installations. There is a potential fire risk with this type of installation.
11. Always ensure that the correct size and type of overcurrent protection is used in motor control circuits. Faulty machines, and inadequately protected associated control circuits, will often burn out long before local overcurrent protection is brought into play.
12. Check that all terminals are tight and secure from mechanical damage. A loose neutral conductor serving a main switch could, under certain fault conditions, be responsible for severe damage caused by a fault current within a protective conductor. This would inevitably lead to its destruction through heat and could possibly affect other current-carrying conductors ranged near or adjacent to the damaged cable.

Fire extinguishers

In the UK, fire extinguishers are colour-coded to enable the user instantly to recognise the type of

TABLE 16.1 British hand-held fire extinguishers (reproduced courtesy Wiltshire Fire Brigade Headquarters).

Colour code	Type	Operational usage	Voltage range (V)
Black	Carbon dioxide	Flammable liquids	+ 1000
Blue	Dry powder	Flammable liquids	up to 1000
Cream	Aqueous film foam	Flammable liquids and solids	Unsafe
Green	Halon 1211 (BCF) (Bromochlorodifluore methane)	Flammable liquids and solids	+ 1000
Red	Water	Solids, paper, cloth and wood	Unsafe

appliance most suited for the emergency in hand. Fumes given off when Halon 1211 (BCF) and carbon dioxide extinguishers are activated may be dangerous if inhaled, especially if used in confined spaces — a point to remember.

Table 16.1 lists five commonly used extinguishers, together with their respective voltage ranges and operational usage.

Handy hints

- Always place a hand-held fire extinguisher where it may be reached quickly.
- Never risk personal injury to fight a fire. Always put your own and your fellow workers safety first. If in doubt, *get out*.
- Never allow a fire to come between you and your route to safety.
- A water-based fire extinguisher must only be used to tackle a fire caused by an electrical fault if the electricity has been turned off at the mains. If in doubt — don't!
- Only attempt to tackle a fire in its earliest stages.

Lifting techniques

Lifting and moving heavy electrical equipment is one of the biggest single causes of personal injury in our industry today. If the load is too heavy, get assistance or employ mechanical means to carry out the task.

As a general rule, loads that are over 20 kg (approximately 44.08 pounds), need to be handled with power-lifting equipment. If you are unable to manage, get help. It is far better to lose a little

pride than spend time recovering from a strained back or a hernia operation.

If several operatives are employed to move a large load, for example a factory-assembled mains control panel, elect one person to provide instructions. This will avoid confusion leading to an unnecessary accident. Should mechanical lifting aids be available and you are trained to use them, take advantage. It will enable the task to be carried out far more easily. Check the load for sharp or ragged edges and wear protective clothing and footwear. Always plan ahead to ensure that the chosen route is free from obstacles and obstructions. Once the load has been safely moved, check that there is sufficient room to place the load with safety.

Points to remember Whenever lifting, keep to this well-tried seven-point procedure illustrated as Figure 16.1:

1. Chin in.
2. Back straight.
3. Elbows tucked into the body.
4. Knees bent.
5. Load gripped firmly. Fingertips should never be used.
6. Lift by use of the leg muscles.
7. Position feet one in front of the other close to the load.

Handy hints

- Always keep bolsters and cold chisels suitably trimmed and sharpened (Figure 16.2). Overhang around the impact area can sheer off causing injury and the sharp edges

3. Never touch the victim with uninsulated hands until removed from the live conductive parts.

4. Do not use any material which might be marginally damp such as a towel or a discarded waste rag.

5. Should the victim be unconscious but breathing and severely burnt, either call the emergency service or drive the patient to your local medical or accident unit — but seek advice.

6. Depending on circumstances, summon assistance if a decision is taken to drive to a local emergency centre.

Coping with an emergency: mouth-to-mouth resuscitation When breathing stops resulting from an accident caused through direct contact with electricity, brain damage will probably occur within three minutes. It is therfore vital that air is forced into the lungs until the casualty is able to commence breathing normally again.

Whenever mouth-to-mouth resuscitation is given, there are seven important stages which must be considered. Familiarise yourself with the procedure and practise with the aid of a medical mannequin.

1. Lay the casualty on his or her back; if practical, use a work bench or table. With head ranged to one side, displace and remove any extraneous material from the mouth (Figure 16.3(a)).

2. Tilt the victim's head fully backwards by placing one hand on the forehead and the other on the neck. (Sniffing the morning air position.) This will enable the mouth to open providing a clear airway passage to the lungs (Figure 16.3(b)). Check that the pulse is present at the neck.

3. If present, transfer the hand from the neck and apply a little upward pressure to the chin. Pinch the nostrils together using your fingers, and blow two rapid but generous breaths by covering the causualty's mouth with your own (Figure 16.3(c)).

4. After the second ventilation, confirm that the chest responds by rising and falling. Should there be no response, tilt the head well back and blow through the victim's nose, closing the mouth with your hand.

5. Continue to breathe steadily into the victim's mouth at a normal breathing rate, monitoring the rise and fall of the chest each time it is artificially inflated.

6. Once breathing is established and is considered normal, recheck the causualty's pulse and place in a recovery position as Figure 16.3(d) illustrates. This will prevent the tongue or body fluids from obstructing the entrance of the windpipe. If the patient is too heavy to move, get assistance.

7. Call the ambulance or if work is carried out on a large construction site, contact the medical centre. Beware of relapse. Stay with your patient until professional medical help arrives.

When the heart has stopped Be calm, be efficient, but above all, *don't panic* — it could save a life. It is advisable to familiarise yourself with the procedure and to seek further advice and guidance from trained professional personnel. Remember — practice leads to perfection!

1. Lay the causualty on his/her back on a flat surface with both legs raised. Clear the air passage by tilting the victim's head fully backwards. Place the head to one side and remove any foreign material from the mouth (Figure 16.4(a)).

2. Confirm that a pulse is present at the neck (the carotid artery). If not, external chest compressions must be started immediately. This is known as *resuscitation*.

3. Locate the centre and bottom of the ribcage. At a distance of two fingers up from this mid-point position, place one hand on top of the other (Figure 16.4(b)), interlocking the fingers of both hands and apply regular and smooth pressure using the heel of the hand. The rib margins in this area are sufficiently flexible to allow up to 30 mm movement.

4. Compress the chest for 15 times at a frequency of between 60 and 80 times per minute. In practical terms this will mean that each chest compression should last between 0.75−1 second.

5. Next, inflate the lungs twice by use of

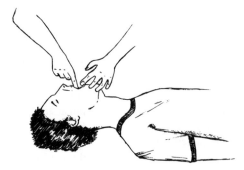

Figure 16.3(a). Place head to one side and remove any extraneous material from the mouth.

Figure 16.3(b). In order to open the air passage, the casualty's head should be tilted fully backwards.

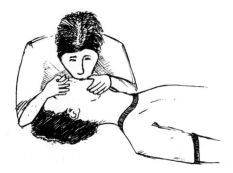

Figure 16.3(c). Pinch the casualty's nostrils together and blow two rapid breaths into the victim's mouth.

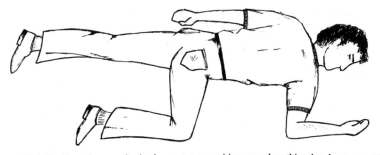

Figure 16.3(d). Place the casualty in the recovery position once breathing has been re-established.

Appendix A   Answers to review and self assessment questions

Part 1

1. Proton, neutron and electron.
2. (d)
3. 1.3 Ω.
4. $\dfrac{R_1}{R_2} = \dfrac{1 + (\alpha t_1)}{1 + (\alpha t_2)}$.
5. (b), (d).
6. Electrons are loosely held.
7. (b).
8. By use of the expression $I = \dfrac{V}{R + r}$ and
   $r = \dfrac{V}{I} - R$.
9. 13 Ω.
10. Extraneous vibration and spikes or anomalies in the supply voltage.
11. To complete the voltage-sensing coil circuit.
12. Series wound; shunt wound.
13. Induction, synchronous and commutator motors.
14. (b).
15. Fans serving air conditioning units.
16. Plates short circuiting.
17. As the potential across a capacitor increases it offers a resistance to current flow. This is known as capacitive reactance.
18. Length, cross-sectional area, temperature and material composition of the conductor.
19. The resistance of a conductor will *decrease* with a rise in temperature.
20. A laminated armature, primary winding and secondary winding.
21. (c).
22. Silver has a large atom with just one valence electron.

23. (c).
24. By use of expressions $I = \dfrac{E}{R}$ and $\dfrac{E_p}{E_s} = \dfrac{I_s}{I_p}$.
25. The electrical resistance offered to a material specimen of unit length and cross-sectional area.

Part 2

1. TT, TN-S, TN-CS and TN-C earthing arrangements.
2. Any conductive part capable of transmitting a potential not forming part of the electrical installation which has a resistance, measured from the main incoming earth terminal, of *less* than 0.25 MΩ.
3. Loop-in or three plate ceiling rose method. Central joint box method. Double pole switch method.
4. Modern kitchens are often high consumers of electricity.
5. Can be rewired. May be installed using mixed sizes of conduit. Can be added to with ease.
6. Gently warm the work-piece.
7. Boiler houses. Fire alarm installations. Emergency lighting systems.
8. It will absorb moisture from the atmosphere (hygroscopic).
9. Clipped to suitable surface or laid directly in the ground.
10. Play a gentle flame around the neck of the element. this method is unsuitable for pre-thermally insulated water storage vessels.
11. Central joint box method.
12. Paper-insulated oil-filled capacitor. Electrolytic capacitor.

13. Electrical non-conductive parts are constructed from laminated sheet iron.
14. Good starting torque but a slight drop in speed with load.
15. Can cause premature wear.
16. 11.

Part 3

1. Carelessness, inexperience, human limitation and horse-play.
2. Injury or direct contact with live components can often be attributed to jewellery being snagged or caught on projecting parts.
3. When working with asbestos or in a dusty environment.
4. If the total hydrogen content of a battery room exceeds 4 per cent, an explosive atmosphere will exist.
5. One.

Answers to self assessment questions

1. Mercury, nickel.
2. Resistance.
3. 5 ohms, 2 amps.
4. 50 ohms, 2 watts
5. $r = \dfrac{V}{I} - R$
6. The type of toroidal transformer and sensing relay used.
7. The slip; usually expressed as a percentage.
8. The phase conductor is connected to *terminal 98* and the neutral conductor is connected to the incoming neutral terminal. The two conductors are then connected to a suitable indicator lamp. It is wise to install an in-line fuse within the phase conductor. This may be accommodated in the starter.
9. 20 μF.
10. Step up transformer. Step down transformer. Isolating transformer.
11. (c), urban development.
12. When a supply authority employs a combined *neutral* and *protective conductor* surrounding an insulated phase conductor to serve an installation.

13. Main gas service. Mains water supply. Oil pipes. Central heating pipes.
14. This technique effectively bonds all extraneous conductive parts together; but not to earth.
15. All cables must be drawn into the conduit together. Keeping the conductors straight and adding powdered french chalk will also help the cables to be drawn smoothly through the conduit.
16. Holes to be no more than one-sixth of the depth of the wooden joist and should be drilled along the neutral axis (Figure 9.25).
17. Exterior installations. Floor screeds. Buried in concrete.
18. Bare copper. PVC covered. LSF (Low Smoke and Fumes) sheathed.
19. (b), twice the nominal working voltage of the installation.
20. Excessive temperature, poor ventilation, dampness, prolonged or frequent use, working voltage of the capacitor rated lower than the voltage applied.
21. The thermal overload relay.
22. Power tools rated at below 0.2 kW, portable hand-held electric drills and small domestic appliances.
23. *British Standards* inspection. Principal visual inspection.
24. Where fixed wiring yields to a current consuming accessory, as for example, a ceiling rose or a socket outlet.
25. (a), 3 months; (b), 1 year (mandatory); (c), 1 year (mandatory); (d), 3 years.
26. (a), False; (b) True; (c) True.
27. Short circuit the conductors and the current protective conductor with an insulated low resistance test resistor after each test.
28. By wearing suitable eye protection.
29. 1. Sound alarm.   2. Try to extinguish the flames.   3. Do not endanger yourself. 4. If the fire cannot be put out, escape by using the safest route out of the building.
30. 1. Chin in.   2. Back straight.   3. Elbows tucked in.   4. Knees bent.   5. Grip load firmly.   6. Lift by use of the leg muscles.   7. Position feet one in front of the other close to the load.

Appendix B Graphical symbols drawn from the International Electro-technical Commission and British Standards 3939

Graphical symbols listed within this appendix (Figure A.1) are used for illustrating electrical installation drawings and plans. The majority conform with recommendeations laid down by the International Electrotechnical Commission (IEC).

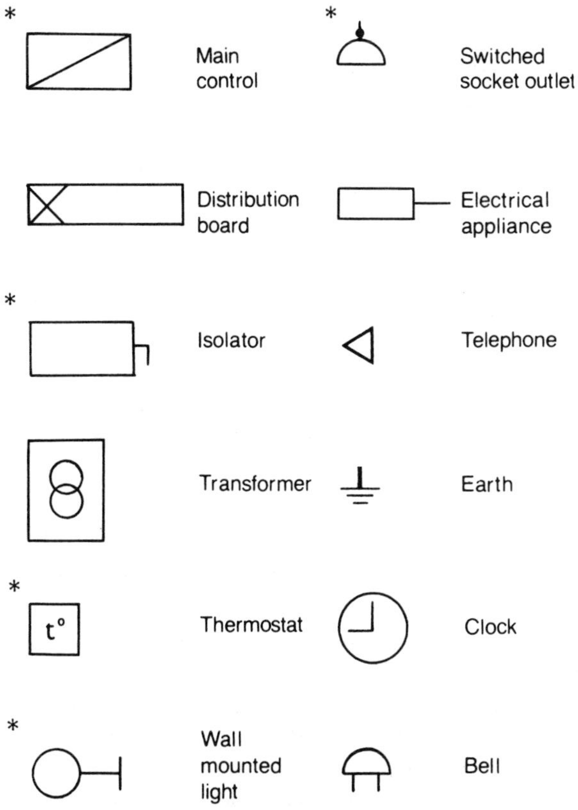

Apart from the United States of America, IEC symbols are widely used throughout the world. Electrographical symbols unadopted by the IEC have been annotated. These have been drawn from British Standards 3939, Section 27, and are in common use in the UK.

Formally accepted symbols and logos should only be used when designing an installation. Fashioning your own may be straightforward and uncomplicated but invariably leads to confusion for others who follow. Regulation 514-09-01 refers.

Graphical symbols drawn from American National Standard, ANSI Y32.9-1972

North American electrical symbols used for installation plans and diagrams differ considerably from those used in Britain. Figure A.2 illustrates a small selection for the purpose of comparison.

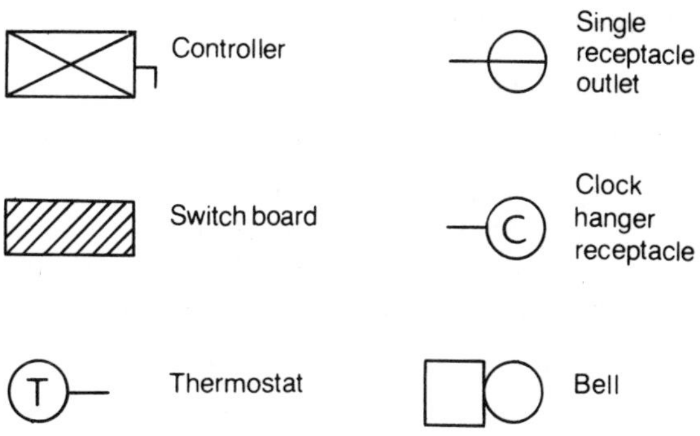

* Main control	* Switched socket outlet
Distribution board	Electrical appliance
* Isolator	Telephone
Transformer	Earth
* Thermostat	Clock
* Wall mounted light	Bell

Figure A.1. Graphical symbols used for architectural and installation plans and diagrams from BS 3939; Section 27, 'Graphical Symbols for Electrical, Power, Telecommunications and Electronic diagrams'.

Controller	Single receptacle outlet
Switch board	Clock hanger receptacle
Thermostat	Bell

Figure A.2. A small selection of North American graphical symbols extracted from ANSI Y 32, 9–1972, 'Graphic Symbols for Electrical Wiring and Layout Diagrams used in Architecture and Building Construction'.

245

Appendix C   Determination of the size of cable required for a given circuit

Undesired heat is constantly being propagated as a by-product whenever current flows within a circuit. A conductor's ability to lose heat energy efficiently will depend on how and where it is installed. An example may be drawn from several current-carrying cables grouped together within trunking where the ability to dissipate heat efficiently would be far less than if the same cables were to be well spaced and clipped to a wall.

High ambient temperature will also hinder the means by which cable heat loss is regulated. It is far more prudent to install a circuit where the ambient temperature is cool than deliberately to choose a route which might be far easier, but where the surrounding temperature is excessive.

On account of this, *correction factors* have been introduced which will, in practice, enable wiring to be carried out using a larger and therefore more suitable cable to serve the circuit under review.

Correction factors

There are four categories which must be considered:

1. C_g Cable grouping factor.
2. C_a Factor for ambient temperature.
3. C_i Factor should the cable be placed within, or is in contact with, thermal insulation:

    (a) Cable contact with one side = 0.75;
    (b) Cable surrounded by insulation = 0.5.

4. C_f A factor of 0.725 is applied where the protective device takes the form of a semi-enclosed, BS 3036, fuse. This factor will not be applicable when the installation is to be carried out using mineral insulated cable.

A circuit served by a semi-enclosed fuse
1. First divide the nominal current of the protective device, I_n, by the correction factor provided for *cable grouping*, C_g. This figure may be found within the IEE Regulations.
2. Next divide the evaluated nominal current of the protective device, I_n, by the correction factor provided for ambient temperature, C_a. Again, this may be located within the IEE Regulations as Tables 4B1, 4B2, 4B3, 4C1 and 4C2.
3. If applicable, now apply the correction factor determined for thermal insulation.
4. Finally further divide by 0.75, the correction factor specified for semi-enclosed fuses.

In practice it is easier to multiply all assembled correction factors together and divide the solution into the nominal current of the protective device, I_n.

As a practical example, consider the following:

A 230 V, 3000 W convector heater, wired using PVC insulated copper conductors, is protected by use of a 15 A semi-enclosed fuse to BS 3036. The installation is planned within an area where the ambient temperature rises to a maximum of 35 °C. The circuit will be grouped and placed within steel conduit accompanied by two other single phase circuits each with identical loads to the proposed convector heater circuit. Determine the minimum size conductor which may be used so as to comply with the IEE Regulations.

1. Mathematically evaluate the proposed design current, I_B, of the circuit. This should be equal to or less than the current rating of the protective device, I_n.

$$I_B = \frac{3000}{230} \qquad [A.1]$$

or, symbolically, $I_B = \frac{W}{V}$

$I_B = 13$ A (this is the design current of the circuit)

2. Next find the correction factor for an ambient temperature of 35 °C (C_a), Regulation Table 4C1. This is shown to be 0.94.
3. Now determine the grouping factor for single core PVC insulated copper cable enclosed in conduit (C_g), Regulation Table 4B1. This is shown to be 0.70.
4. Finally, apply the correction factor for a semi-enclosed BS 3036 fuse (C_f). This was given to be 0.725.

Evaluating the minimum current carrying capacity of the cable to be used for the proposed circuit:

$$I_z = \frac{I_n}{C_a \times C_g \times C_f} \qquad [A.2]$$

where I_z is the current-carrying capacity of the conductor,
I_n is the current rating of the protective device,
C_g is the correction factor for cable grouping,
C_a is the correction factor for ambient temperature and
C_f is the correction factor given for semi-enclosed fuses.

Evaluating Expression [A.2]:

$$I_z = \frac{15}{0.94 \times 0.7 \times 0.725}$$

$$I_z = \frac{15}{0.476}$$

$$I_z = 31.5 \text{ A}$$

A cable may now be selected from Table 4A of the IEE Regulations under 'Installation method 3'. This will show that a 4 mm² conductor having a current-carrying capacity of 32 A will meet the requirements of this regulation in order to serve the proposed circuit. Reference is made to Table 4D1A.

Other forms of protective device
Should the protective device take the form of:

1. fuse to BS 88 or BS 1361,
2. circuit breaker to BS 3871, Part 2,
3. circuit breaker to BS 4572, Part 2,

the following expression may be used to calculate the current-carrying capacity of the proposed conductor, I_z.

$$I_z = \frac{I_n}{C_a \times C_g \times C_i} \qquad [A.3]$$

This is similar to Expression [A.2] but the correction factor C_f has been omitted.

Volt drop
It is also important to consider volt drop in any calculation made to determine the size of conductor to be used. This has been reviewed in Chapter 6. Appendices 4 and 7 of the *16th Edition IEE Wiring Regulations* offer a selection of values of voltage drop in the form of simple, easy-to-read tables.

Appendix D   Visual identification of input services to buildings

British Standards directive, 1710: 1984 (1989) provided means for identifying pipe line and service conduits installed in buildings and on board ships.

Each individual service is allotted a basic colour to provide means of identification and is fashioned as a band of coloured self-adhesive tape some 150 mm wide which is stuck to the wall or cladding of the service pipe.

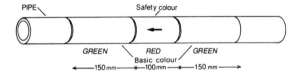

Figure A.3. The contents of this water pipe is used for fire fighting.

If it is essential that the exact contents of the pipe are known, a safety or reference coloured band approximately 100 mm in width is sandwiched between two 150 mm printed vinyl-based coloured bands as illustrated in Figure A.3.

Conduit containing low voltage electrical mains cables is painted using the secondary colour *orange* (BS colour reference number 06 E 51), as demanded by *Regulation 514-02-01*.

Table A1 provides a selection of typical services found in a modern industrial establishment, whereas Table A2 offers informative and safety details of the service pipes concerned.

TABLE A1   Services to buildings; means of identification

Service	Basic colour identification	BS colour reference
Acids and Alkalis	Violet	22 C 37
Air	Light blue	20 E 51
Electricity	Orange	06 E 51
Fresh water	Auxiliary blue	18 E 53
Fuel oil	Brown	06 C 39
Gas	Yellow ochre	08 C 35
Other liquids	Black	00 E 53
Steam	Silver grey	10 A 03
Vacuum	Light blue	20 E 51

TABLE A2   Service pipe information — contents advice.

Information/usage	Safety colour	BS colour reference
Warning	Yellow	08 E 51
Danger	Yellow and black stripes	Yellow 1
Fresh water	Auxiliary blue	18 E 53
Fire fighting	Red	04 E 53

Appendix E   NVQ study guide

The table below correlates NVQ performance criteria and knowledge requirements to the relevant page and Chapter numbers in this book. 'AEI' indicates that the subject is discussed in this book's companion volume *Advanced Electrical Installations*, also published by Longman.

Level 1   Fabricating and fixing electrical cable supports

Unit		Page or Chapter numbers
1C11b	Prepare, maintain and restore work sites for fixing cable supports	87, 97, 114, 115, 215–17, 225–7, 233, Ch. 8 & 10, Appendix D, AEI
1C11c	Install and dismantle access platforms	225–31
1C12b	Establish and maintain effective working relationships	87, 88, Part 3, AEI
1C13b	Maintain the healthy and safe working environment	85, 86, 140, 237, 240, Part 3
1C21b	Fix cable supports	85, 113, 125–33, 141, 142, 144–7, 215, 220, 223, 225–31, Ch. 9 & 10, Appendix D, AEI
1C21e	Cut and shape cable supports	125–8, 130–33, 140, 144, 147, Ch. 10, AEI

Level 2   Installing electrical systems and equipment

Unit		Page or Chapter numbers
2C11c	Prepare, maintain and restore the work site for electrical installation	139, 190, 215, 217, 218, 225, 226, 233, Ch. 15, 248, AEI
2C12c	Establish and maintain effective working relationships	87, AEI
2C13c	Maintain the healthy and safe working environment	240, Ch. 15 & 16, AEI
2C21c	Install components of electrical installations	97, 119, 175–84, 226, Ch. 4, 6, 8, 9, 10, 11, 12, 13 & 14, AEI
2C22c	Confirm the installation for commissioning	187, 190, Ch. 14
2F43c	Contribute to the improvement of the organisation's services to its customers	AEI

Level 3 Installing and commissioning electrical systems and equipment

Unit		Page or Chapter numbers
3B11d	Contribute to electrical design objectives	Ch. 15 & 16, AEI
3B21d	Review the site characteristics for installation and services	216, 227, Ch. 7, 10, 11 & 14, AEI
3C11d	Prepare the work site for electrical installation. Maintain and reinstate	190, 215, 218, 227, AEI
3C12d	Create and preserve effective working relationships	Ch. 15 & 16, AEI
3C13d	Work site health and safety	Ch. 15 & 16, AEI
3C21d	Install, fix and test electrical installations	18–25, 76, 206, 226, Part 1, Ch. 7, 8, 11, 13 & 14, Appendix D, AEI
3C22d	Commission electrical installations	76, 197, Ch. 8, 10, 11 & 14, AEI
3C32d	Identify and make good fault conditions	22, 76, 206, 227, Part 1, Ch. 6, 10, 11, 12 & 14
3D11d	Comply with statutory and regulatory requirements	217, 220, 225, 233, 245, Ch. 14, AEI
3D21d	Monitoring progress of installation and operating costs	AEI
3F43d	Contribute to the improvement of your company's services to their customers	AEI

Index